Run-time Adaptation for Reconfigurable Embedded Processors

T0180926

Lars Bauer · Jörg Henkel

Run-time Adaptation for Reconfigurable Embedded Processors

Springer

Lars Bauer
Karlsruhe Institute of Technology
Haid-und-Neu-Str. 7
76131 Karlsruhe
Germany
lars.bauer@kit.edu

Jörg Henkel
Karlsruhe Institute of Technology
Haid-und-Neu-Str. 7
76131 Karlsruhe
Germany

ISBN 978-1-4899-8199-8   ISBN 978-1-4419-7412-9 (eBook)
DOI 10.1007/978-1-4419-7412-9
Springer New York Dordrecht Heidelberg London

Printed on acid-free paper

Springer is part of Springer Science+Business Media (www.springer.com)

Contents

Abbreviations

AC   Atom container: a part of the reconfigurable fabric that can be dynami-
            cally reconfigured to contain an atom, i.e. an elementary data path
AGU   Address generation unit
ALU   Arithmetic logic unit
ASF   Avoid software first: a reconfiguration-sequence scheduling algorithm,
            as presented in Sect. 4.5
ASIC   Application-specific integrated circuit
ASIP   Application-specific instruction set processor

BC   Bus connector: connecting an → AC to the atom infrastructure
BRAM   Block → RAM: an on-chip memory block that is available on
            Virtex → FPGAs

cISA   core Instruction Set Architecture: the part of the instruction set that is
            implemented using the (nonreconfigurable) core pipeline; can be used
            to implement → SIs as well, as presented in Sect. 5.2
CLB   Configurable logic block: part of an → FPGA, contains
            multiple → LUTs
CPU   Central processing unit

DCT   Discrete cosine transformation: a computational kernel that is used in
            H-264 video encoder

EEPROM   Electrically erasable programmable read only memory

FB   Forecast block: indicated by an → FI, containing a set of → SIs with
            a → FV per SI
FI   Forecast instruction: a special → HI that indicates an → FB
FIFO   First-in first-out buffer
FPGA   Field programmable gate array: a reconfigurable device that is com-
            posed as an array of → CLBs, → BRAMs, and further components
FPS   Frames per second

FSFR	First select, first reconfigure: a reconfiguration-sequence scheduling algorithm, as presented in Sect. 4.5
FSL	Fast simplex link: a special communication mechanism for a MicroBlaze processor
FSM	Finite state machine
FV	Forecast value: the expected number of $\rightarrow$ SI executions for the next computational block, part of the information in an $\rightarrow$ FB
GPP	General purpose processor
GPR	General purpose register file
GUI	Graphical user interface
HEF	Highest efficiency first: reconfiguration sequence-scheduling algorithm, as presented in Sect. 4.5
HI	Helper instruction: an assembly instruction that is dedicated to system support (e.g., an $\rightarrow$ FI); not part of the $\rightarrow$ cISA and not an $\rightarrow$ SI
HT	Hadamard transformation: a computational kernel that is used in H-264 video encoder
IP	Intellectual property
ISA	Instruction set architecture
ISS	Instruction set simulator
KB	Kilo byte (also KByte): 1,024 byte
LSU	Load/store unit
LUT	Look-up table: smallest element in an $\rightarrow$ FPGA, part of a $\rightarrow$ CLB; configurable as logic or memory
MB	Mega byte (also MByte): 1,024 $\rightarrow$ KB
MinDeg	Minimum degradation: an atom replacement algorithm, as presented in Sect. 4.6
MUX	Multiplexer
NOP	No operation: an assembly instruction that does not perform any visible calculation, memory access, or register manipulation
NP	Nondeterministic polynomial: a complexity class that contains all decision problems that can be solved by a nondeterministic Turing machine in polynomial time
OS	Operating system
PCB	Printed circuit board
PRM	Partially reconfigurable module
PSM	Programmable switching matrix

RAM   Random access memory

RFU   Reconfigurable functional unit: denotes a reconfigurable region that
           can be reconfigured toward an SI implementation

RISPP   Rotating instruction set processing platform

SI   Special instruction

SJF   Shortest job first: a reconfiguration sequence-scheduling algorithm, as
           presented in Sect. 4.5

SPARC   Scalable processor architecture: processor family from Sun
           Microsystems; used for the → RISPP prototype

VLC   Variable length coding: a computational kernel that is used in H-264
           video encoder

VLCW   Very long control word: configuration bits for the coarse-grained
           reconfigurable atom infrastructure, i.e. determining the operation mode
           and connection of the → ACs, → BC, → AGUs, and → LSUs

List of Figures

List of Tables

List of Tables

List of Algorithms

Chapter 1
Introduction

Embedded processors are the heart of embedded systems. They perform the calculations that are required to make a set of components eventually appear as a system. For instance, they receive input data from cameras, microphones, and/or many further sensors, perform the calculations that the user of the embedded system expects (e.g., video compression/transmission, feature extraction, etc., i.e., the actual functionality of the system) and present the corresponding results. Therefore, embedded processors are the key components for rapidly growing application fields ranging from automotive to personal mobile communication/entertainment, etc.

Designing an embedded system typically starts with an analysis of the requirements. Afterwards, some standard components (e.g., periphery or processors provided as chips or IP cores) may be selected and combined with some specifically created components to realize the desired system either as a single chip or as a board with multiple chips. For some tasks that the system shall be able to perform (e.g., decoding a video stream) the question arises whether a general-purpose processor (GPP) or an application-specific integrated circuit (ASIC) shall perform that task. Choosing a GPP has the advantage that it provides a very high flexibility, i.e., it is programmable by software. This allows that the GPP can be used to perform further tasks (when the video decoder is not required at a certain time or in parallel to it, using a multitasking environment). In addition, it also allows changing the specification of the tasks (e.g., a new video decoding standard or a bug fix) by changing the software, e.g., via firmware upgrades. The advantage of an ASIC is its efficiency (e.g., "performance per area" or "performance per power consumption") that it achieves because it is particularly optimized for the specific task. Using an ASIC typically introduces rather high initial cost for creating that ASIC, but the higher efficiency may overcome the initial monetary drawback. However, an ASIC does not provide flexibility, i.e., it cannot be used to perform any tasks that were not considered when designing it, and changes of the initial task specification/implementation are not possible without a rather costly redesign. Depending on the projected work area (e.g., selling volume, required possibility for upgrades, reusability for further products, etc.), GPPs or ASICs may be used.

L. Bauer and J. Henkel, *Run-time Adaptation for Reconfigurable Embedded Processors*,
DOI 10.1007/978-1-4419-7412-9_1, © Springer Science+Business Media, LLC 2011

1.1   Application-Specific Instruction Set Processors

The term "ASIP" (application-specific instruction set processor) was introduced in
the early 1990s and denoted then processors that are application specific. They provide
an alternative to GPPs and ASICs when designing an embedded system as they – to
some degree – combine both approaches. Designing an ASIP involves analyzing
(e.g., profiling) an application, a set of applications, or even a whole application
domain (e.g., multimedia) and then creating an instruction set that performs very
efficient for that application domain compared to a GPP. The implementation of this
instruction set extension correspond to ASIC parts that are dedicated to a particular
application and that are embedded into to core of a GPP. The application can use
these ASIC parts by so-called special instructions (SIs). An ASIP may provide the
same functionality as a GPP, but depending on the optimization targets, also parts of
the GPP functionality (e.g., a floating-point unit) may be removed, thus potentially
reducing the flexibility. Altogether, an ASIP provides a trade-off between GPPs and
ASICs. Since the late 1990s/early 2000s, the term ASIP is far more expanded. Since
major vendors like Tensilica [Tena], ARC [ARC], CoWare [CoW], etc. offer their
tool suites and processors cores, the user can now create a very specific instruction
set (instead of buying/licensing a predetermined one) that is tailor-made for a certain
application. Typically, these tool suites come with a whole set of retargetable tools
such that code can easily be generated for that specific extensible processor. As a
result, extensible processors are more efficient than the first generation of ASIPs.

Nowadays, the landscape of embedded applications is rapidly changing and it can
be observed that today's embedded applications are far more complex and offer a far
wider set of functionality than a decade ago. This makes it increasingly difficult to
estimate a system's behavior sufficiently accurate at design time. In fact, after exploration of complex real-world embedded applications (most of them from the multimedia domain) it became apparent that it is hard or even impossible to predict the
performance and other criteria accurately during design time. Consequently, the more
critical decisions are fixed during design time, the less flexible embedded processors
can react to nonpredictable application behavior. This not only results in a reduced
efficiency but also leads to an unsatisfactory behavior when it comes to the absolute
design criteria "performance" and "power consumption." However, the ASIP
approach assumes that customizations are undertaken during design time with little
or no adaptation possible during run time. In addition, for large applications that
feature many diverse computational hot spots and not just a few exposed ones, current
ASIP concepts struggle. According our studies with truly large and inherently diverse
applications, customization for many hot spots introduces a nonnegligible overhead
and may significantly bloat the initial small processor core, because rather many different SIs need to be provided and implemented. Then, a designer needs to answer
the question whether more powerful processor architectures would not have been the
better choice. Since often only one hot spot is executed at a certain time, the major
hardware resources reserved for other hot spots are idling. This indicates an ineffi-
ciency that is an implication of the extensible processor paradigm.

1.2 Reconfigurable Processors

Reconfigurable computing provides adaptivity by using a fabric that can be reconfigured at run time (see Sect. 2.2). This reconfigurable fabric can be used to implement application-specific accelerators, similar to those that are used by ASIPs. Great efforts were spend in investigating how such a reconfigurable fabric can be connected with a processor and used by an application efficiently. Recent approaches allow coupling the reconfigurable fabric into to the pipeline of the processor as reconfigurable functional units (RFUs). They allow providing implementations of multiple SIs (like the SIs that are developed and used for ASIPs), by loading reconfigurable accelerators into the RFUs. However, the RFUs provide the conceptual advantage that they can be reconfigured and thus it is not predetermined which SI implementations shall be available in the RFUs at which time. Therefore, such a reconfigurable processor is no longer specific for a particular application or application domain. Even after the reconfigurable processor is fabricated and deployed, the reconfigurable accelerators can be modified (similar to the above-discussed firmware update) to support extended application standards and even new application domains. This is done by providing modified or additional configuration data that is used to reconfigure the RFUs. This increases the flexibility in comparison to ASIPs as the software and parts of the hardware can be changed at run time. In addition, the application-specific reconfigurable accelerators provide the efficiency of an ASIC implementation. Therefore, reconfigurable processors present a promising trade-off between ASICs and GPPs that goes beyond the concept of ASIPs.

However, the process of reconfiguration also comes with some drawbacks. Performing a reconfiguration is a rather slow process (multiple milliseconds) and thus the potential benefit of run-time reconfiguration may be diminished. If an application requires rather few SIs and all of them fit into the RFUs at the same time, then the reconfiguration overhead only occurs when the application starts, i.e., all accelerators are reconfigured into the RFUs once. In this case, the reconfiguration overhead will typically amortize over time. However, if a more complex application demands more SIs than fit into the RFUs at the same time, then frequent run-time reconfigurations of the RFUs are required to exploit the performance of a reconfigurable processor maximally. In such a case, the overall application performance may be affected by the reconfiguration time. Actually, when an SI implementation provides more parallelism, then this leads to larger area requirements and the reconfiguration time becomes correspondingly longer. In general, a trade-off between provided parallelism and reconfiguration overhead needs to be determined. Here, the performance of an ASIP is not bound by the parallelism of an SI implementation, but rather its area efficiency is bound by it, because SI implementations that provide more parallelism leads to a larger ASIP footprint.

Another important aspect of reconfigurable processors is that, even though the hardware can be reconfigured at run time, typically it is determined at compile time "what" shall be reconfigured and "when" the reconfiguration shall be performed. Certainly, this limits the run-time adaptivity, as the deployment of the run-time adaptive RFUs is predetermined at compile time. However, run-time adaptivity is important to use the RFUs most efficiently. For instance, our exploration of real-world applications points out that the execution time of some computational hot spots may highly depend on input data. As this input data (e.g., an input video from camera) is not known at compile time, it cannot be predicted, which out of multiple hot spots in an application will be executed for how long? However, this information affects, which application-specific accelerators will provide which performance improvement for the overall application execution. Still, the decision which accelerators shall be reconfigured to the RFUs is determined at compile time, i.e., without the demanded run-time knowledge. Given that not all accelerators fit to the RFUs at the same time, the full performance potential cannot be exploited. Especially in multitasking scenarios, it is typical that not all demanded accelerators fit to the RFUs, because they need to be shared among the executing tasks. Often it is not even known at compile time, which applications will execute at the same time. For instance, if a user downloads and starts additional applications on his cell phone or handheld device, then it is impossible to consider all potential execution scenarios when compiling a particular application. Therefore, it is not known which share of the RFUs is available for a particular application and thus, the design and implementation of SIs cannot be optimized for it. A concept and strategy to shift these decisions to a run-time system that is aware of the run-time specific scenario (i.e., which applications execute and which input data they process) can improve the efficiency of run-time reconfigurable processors significantly.

1.2.1 Summary of Reconfigurable Processors

Performance and efficiency are key targets already for today's embedded systems and thus in particular for embedded processors. Reconfigurable processors can be adapted to different applications and application domains after fabrication without redesign, which provides a fast time-to-market and reduced nonrecurring engineering cost in comparison to ASIPs. However, their performance may be limited by the reconfiguration overhead and – due to compile-time-determined reconfiguration decisions – they do not exploit their full potential for run-time adaptivity. Especially compile-time unpredictable scenarios would benefit from improved adaptivity. For instance, the execution time of a computational hot spot may depend on input data, and the utilization of the RFUs depends on the number and priority of the executing tasks. These are the major challenges for future embedded processors that have to face rather complex applications, various application domains, and compile-time unpredictable scenarios, e.g., in cell phones or handheld devices.

1.3 Contribution of this Monograph

The aim of the work presented in this monograph is to improve the adaptivity and efficiency of state-of-the-art embedded processors to provide means for challenging and complex next-generation applications. This monograph presents the novel rotating instruction set processing platform (RISPP) that gives origin to an innovative thinking process and moves traditional design-/compile-time jobs to run time, thus providing adaptivity in order to increase the efficiency and performance. It uses the available hardware resources efficiently by applying the novel concept of *modular SIs* that is explained later on. Compared to state-of-the-art reconfigurable processors, this reduces the time until a certain SI is accelerated by the reconfigurable fabric. This is achieved by the ability to utilize elementary reconfigurable data paths without the constraint to wait until the complete reconfiguration of an SI (composed out of multiple data paths) is completed. In the proposed modular SI composition, an SI is composed of elementary data paths as a connected module, which is mainly driven by the idea of a high degree of reusability of data path elements.

In particular, the novel contributions presented in this monograph are as follows:

- *The novel concept of modular special instructions.* This enables a *dynamic* and *efficient* trade-off between different "parallelism vs. reconfiguration overhead" implementation alternatives, which solves the before-mentioned problem that providing more parallelism may actually lead to a slower execution time due to increased reconfiguration overhead. "Dynamic" means that the trade-off can be adapted at run time, depending on the application requirements or the particular situation in a multitasking environment. "Efficient" means that the implementation alternative can be upgraded incrementally. For instance, if an SI implementation shall be adapted to provide a more parallel version, then it is not required to load the more parallel version from scratch, but instead the currently available implementation can be upgraded (i.e., only the additionally required data paths need to be reconfigured).
- *A novel run-time system to determine the reconfigurations.* To utilize the unique features of modular SIs and to provide a very high adaptivity (as demanded by applications with high input-data-dependent control flow and by multitasking systems with unpredictable user interactions) a run-time system is required to determine the reconfiguration decisions. In particular, this monograph proposes the following:

  - An SI online monitoring and prediction scheme that is based on an error back propagation.
  - A novel run-time SI implementation selector that chooses an SI implementation dynamically, considering the trade-off between area and performance (state-of-the-art reconfigurable processors instead statically offer one SI implementation and predetermine when it shall be reconfigured).

- A novel run-time data path reconfiguration scheduler that determines a performance-wise advantageous reconfiguration sequences for data paths and thus determines the SI upgrades.
- A novel performance-guided replacement policy that is optimized for the requirements of reconfigurable processors that need to replace reconfigurable data paths at run time.

- *A novel hardware architecture.* This introduces a novel infrastructure for computation and communication that actually enables the implementation of modular SIs, allows SI upgrading by stepwise performing further reconfigurations toward a more parallel implementation, and offers different design-time parameters to match specific requirements.

Furthermore, this monograph provides a formal problem description for modular SIs and for all algorithms of the run-time system. Detailed analysis and evaluations are presented for the individual algorithms of the run-time system and the hardware architecture. In addition, challenging comparisons with state-of-the-art ASIPs and reconfigurable processors demonstrate the superiority of the presented approach when facing a complex H.264 video encoder application. In addition to a flexible and parametrizable simulation framework that was developed in the scope of the work presented in this monograph, the presented work is also implemented in a hardware prototype that is based on an FPGA to perform partial run-time reconfigurations in practice.

The adaptive extensible RISPP processor that is presented in this monograph improves upon state-of-the-art approaches due to its novel vision of modular SIs and a run-time system that uses the modular SIs to enable an adaptive and efficient utilization of the available reconfigurable fabric without statically predetermined reconfiguration decisions. These novel concepts allow utilizing the available hardware resources efficiently on an as-soon-as-available basis. That means as soon as a data path is reconfigured, it may be used to compose the functionality of SIs. Over time, the SIs may then be gradually upgraded to full performance. In what specific composition an SI is available at a certain point in time is not known at compile time since it depends on the context during run time.

1.4 Monograph Outline

Before presenting the actual contribution of this monograph, *Chap. 2* presents background for extensible processors and different techniques and approaches for reconfigurable computing before discussing state-of-the-art related reconfigurable processors.

Chapter 3 analyzes the problems of state-of-the-art reconfigurable processors in detail and then presents the foundation for the solution that is presented in this monograph, i.e., modular special instructions (SIs). After discussing how they conceptually diminish or solve the problems of related reconfigurable processors while furthermore providing extended adaptivity, several real-world examples for modular SIs are presented and a formal description is introduced together with functions that operate on it. This formal description is used in the next chapter to describe the tasks of the run-time system and the developed pseudo codes in a clear and precise manner.

In *Chap. 4*, the key novel contribution is presented in detail, i.e., the run-time system that eventually exploits the adaptivity potential that is provided by the proposed modular SIs. At first, a short overview of the developed architecture is given, which is required to describe the run-time system details. Afterwards, an analysis of the requirements for the run-time system (e.g., which operations need to be performed at run time and to achieve adaptivity, which may be performed during compile time) is presented together with a first description of its components. Subsequently, the individual components are presented in different sections, i.e., online monitoring to predict the SI execution frequency, selecting an SI implementation, scheduling the corresponding reconfigurations, and deciding the demanded replacements. Each component is modeled on a formal basis, described in detail (using pseudo codes and examples), evaluated, and benchmarked (e.g., comparing different parameter settings or algorithmic alternatives).

The hardware architecture to realize the envisioned novel concepts is presented in *Chap. 5*. After describing the SI instruction format details and the implementation of a trap handler to realize the SI functionality when the reconfigurations are not completed yet, the data memory access and especially the novel hardware framework that realizes modular SIs are presented in detail and evaluated for different parameters. Eventually, the chapter presents the implementation results for the developed and tested hardware prototype including performance results for the algorithms of the run-time system executing on the prototype.

After benchmark results for the run-time system and the hardware implementation were already presented in Chaps. 4 and 5, *Chap. 6* presents an evaluation for the entire RISPP approach, facing different architectural parameters (e.g., provided reconfigurable fabric and memory bandwidth, etc.) and comparing the RISPP approach with state-of-the-art ASIPs and reconfigurable processors from literature.

Eventually, *Chap. 7* concludes this monograph and provides an outlook, *Appendix A* presents the simulation environment that was developed and used in the work described in this monograph, and *Appendix B* shows details for the FPGA board and the developed printed circuit board (PCB) that were used to implement the RISPP hardware prototype.

Chapter 2
Background and Related Work

Embedded processors are the key for rapidly growing application fields ranging from automotive to personal mobile communication, computation, entertainment, etc. The work presented in this monograph envisions an embedded processor that follows the concepts of extensible processors and uses a reconfigurable fabric to provide application-specific accelerators. Even though this work is mainly related to run-time reconfigurable processors, this chapter presents a short overview about the concepts of extensible processors and provides general background for reconfigurable computing before reviewing the most prominent work from the area of reconfigurable processors with a focus on state-of-the-art processors with a run-time reconfigurable instruction set.

2.1 Extensible Processors

In the early 1990s, the term ASIP has emerged denoting processors with an application-specific instruction set, i.e., they are specialized toward a certain application domain. They present a far better efficiency in terms of "performance per area," "performance per power," etc. compared to mainstream processors and eventually make today's embedded (and often mobile) devices possible. The term ASIP comprises nowadays a far larger variety of embedded processors allowing for customization in various ways including (a) instruction set extensions, (b) parameterization, and (c) inclusion/exclusion of predefined blocks tailored to specific applications (like, e.g., an MPEG-4 decoder) [Hen03]. A generic design flow of an embedded processor can be described as follows:

1. An application is analyzed/profiled.
2. An extensible instruction set is defined.
3. The extensible instruction set is synthesized together with the core Instruction Set Architecture (cISA).
4. Retargetable tools for compilation, instruction set simulation, etc., are (often automatically) created.

L. Bauer and J. Henkel, *Run-time Adaptation for Reconfigurable Embedded Processors*, DOI 10.1007/978-1-4419-7412-9_2, © Springer Science+Business Media, LLC 2011

5. The application characteristics (e.g., area, performance, etc.) are analyzed.
6 The process might be iterated several times until design constraints comply.

A general overview of the benefits and challenges of ASIPs is given in [Hen03, KMN02]. Tool suites and architectural IPs for embedded customizable processors with different flavors are provided by major vendors like Tensilica [Tena], CoWare/ LisaTek [CoW], ASIP Solutions [ASI], ARC [ARC], and Target [Tar]. In addition, academic approaches like PEAS-III [IHT+00, KMTI03], LISA [HKN+01, ISS], and Expression [HGG+99] are available.

Using these tools suites, the designer can now implement a specific instruction set that is tailor-made for a certain set of applications. Typically, these suites come with a whole set of retargetable tools (compiler, assembler, simulator, debugger, etc.) such that the code can be generated conveniently for a specific ASIP. As the instruction set definition requires both, application and hardware architecture expertise, major research effort was spent in design-space exploration [CAK+07] and automatically detecting and generating so-called special instructions (SIs) from the application code [CZM03, HSM07]. A library of reusable functions (manually designed but therefore of high quality) is used in [CHP03], whereas in [API03, SRRJ03] the authors describe methods to generate SIs from matching profiling patterns. The authors in [BCA+04] investigate local memories in the functional units, which are then exploited by SIs. An automated, compiler-directed system for synthesizing accelerators for multiple loops (multifunction loop accelerators) is presented in [FKPM06]. The authors in [BKS04] present an approach to exploit similarities in data paths by finding the longest common subsequence of multiple data paths to increase their reusability. [CPH04] introduces an estimation model for area, overhead, latency, and power consumption under a wide range of customization parameters. In [BNS+04] an approach for an instruction set description on architecture level is proposed, which avoids inconsistencies between compiler and instruction set simulator.

However, facing the requirements of modern embedded systems, ASIPs experience a practical limitation. These approaches assume that customizations are undertaken during design time with little or no adaptation possible during run time. Therefore, they may perform poorly when deployed in scenarios that were not considered during optimizations. The proposed RISPP approach combines the paradigms of extensible processor design with the paradigm of dynamic reconfiguration in order to address the following concerns in embedded processing:

1. An application might have many computational hot spots (instead of only a few) and would require a large additional chip area in order to comprise all customizations necessary.
2. The characteristics of an application may widely vary during run time due to switching to different operation modes, change in design constraints (systems runs out of energy, for example), or highly uncorrelated input stimuli patterns.

In addition, for large applications that feature many hot spots and not just a few exposed ones, current ASIP concepts struggle. In fact, customization for many hot spots introduces a nonnegligible overhead and may bloat the initial small processor core.

Then, a designer needs to answer the question whether more powerful processor architectures would not have been the better choice. One means to address this dilemma is reconfigurable computing since resources may be utilized in a time-multiplexed manner (i.e., reconfigured over time), thus combining the performance and efficiency of dedicated hardware accelerators with a flexibility that goes beyond that of ASIPs and ASICs, respectively.

2.2 Reconfigurable Processors

The reconfigurable instruction set processing platform (RISPP) that is presented in this monograph utilizes the techniques of reconfigurable computing. Reconfigurable architectures address the challenge of supporting many hot spots by reusing the available hardware in time-multiplex, i.e., reconfiguring its functionality to support the currently executed hot spots. In this section, general techniques and concepts are presented, before reviewing state-of-the-art projects that are related to the proposed RISPP approach. General overviews and surveys for reconfigurable computing can be found in [Ama06, BL00, Bob07, CH02, Har01, HM09, TCW+05, VS07]. They also cover older projects and areas that are of less relevance for the approach that is presented in this monograph.

2.2.1 Granularity of the Reconfigurable Fabric

Conceptually, reconfigurable architectures can be separated into coarse and fine grained [VS07]. The coarse-grained approach maps word-level computation to a configuration of an array of arithmetic logic units (ALUs), e.g., using an automatic framework to select the appropriate configurations [HSM07]. The fine-grained approach instead reconfigures look-up tables (LUTs) on bit level (e.g., field programmable gate arrays: FPGAs). Therefore, the major difference between fine- and coarse-grained reconfigurable fabrics is the granularity of the reconfigurable elements. For coarse-grained fabrics, the operation of a word-level ALU (typically between 16 and 32 bit) can be determined with configuration bits. For fine-grained fabrics, the operation of a bit-level LUT (typically between $6:1^1$ and $4:1$) can be determined with configuration bits. To implement a control- and/or data-flow, multiple configurable elements (ALUs or LUTs, respectively) are connected (the connections are determined by configuration bits as well). Conceptually, word-level computation can be mapped to coarse-grained reconfigurable fabrics rather efficient and byte or sub-byte-level computation performs more efficient on fine-grained reconfigurable fabrics. For instance, fine-grained reconfigurable fabrics are

---

[1] That is, 6-bit input and 1-bit output.

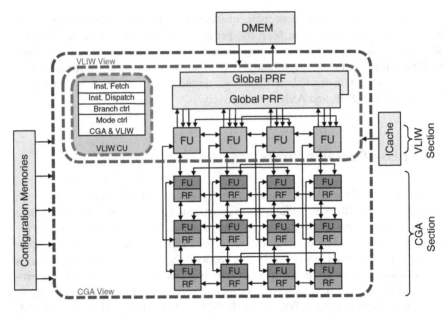

Fig. 2.1 Connecting coarse-grained reconfigurable functional units

Legend:

CGA:	Coarse-Grained Array	FU:	Functional Unit
CU:	Control Unit	PRF:	Predicate Register File
DMEM:	Data Memory	RF:	Register File
DRF:	Data Register File	VLIW:	Very Long Instruction Word

well-suited for diverse operations on image pixels (byte level), implementing state machines (sub-byte level), or calculating conditions (bit level). In addition to fine- and coarse-grained reconfigurable architectures, also combinations of both kinds were investigated [ITI, KBS+10, TKB+07].

Figure 2.1 shows a particular instance of the ADRES [BBKG07, MVV+03] coarse-grained reconfigurable fabric. The array comprises coarse-grained reconfigurable functional units (FUs). In general, an FU may correspond to an ALU; however, typically more complex compositions are used. For instance, it may comprise multiple ALUs or an ALU and Adder, etc. Furthermore, each FU may contain a local register file and also heterogeneous FUs may be used (e.g., specialized FUs for multiplication). The FUs are connected with each other. Often a two-dimensional arrangement is used and the FUs are connected to their direct neighbors (sometimes also to farther distant FUs and sometimes – when the data may through the array is predetermined – only to some neighbors). Different FU compositions, connections, heterogeneity, and memory connections (local register file and/or data memory) may be used for coarse-grained reconfigurable architectures. ADRES investigates different architecture instantiations [BBKG07, MLV+05] with a special focus on an automatically generated compiler that can map kernels from C Code to the ADRES array [MVV+02].

A different approach for the coarse-grained reconfigurable computing is the custom compute accelerator (CCA) [CBC+05, CKP+04]. The CCA approach is based on coarse-grained reconfigurable elements and it supports a run-time placement of elementary data-flow graphs. This means, instead of preparing the configuration of the reconfigurable array at compile time (as done by ADRES), CCA identifies relevant computational kernels at compile time, but it creates configurations to accelerate them at run time. This allows that one application may be accelerated by different instantiations of the CCA reconfigurable array. To simplify the task of run-time placement, CCA is limited to a significantly narrowed hardware architecture. The reconfigurable array is meant to realize straight word-level data-flow (no external memory access, no internal state registers, and no data feedback, i.e. only acyclic data-flow graphs). As these straight data flows typically have more input data than produced output data (thus forming a triangular data-flow graph), the authors have organized their coarse-grained elements in a triangular shape as well. As shown in Fig. 2.2, the specific size, shape, and interconnections are determined application- and domain specific, depending on the occurring data-flow graphs. Placing individual nodes of a data-flow graph in a CCA can then be accomplished with one pass over the operations in the graph by placing each node

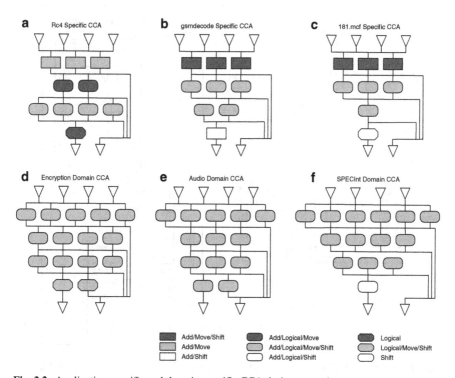

Fig. 2.2 Application-specific and domain-specific CCA design examples

in the highest row that can support the operation while respecting data dependencies. Therefore, this concept cannot be used for complex data-flow graphs (e.g., with embedded control-flow or exploiting loop-level parallelism), computations with high memory requirements (typical for multimedia applications), and multiple parallel executing applications with diverse system requirements (i.e., diverse-shaped data-flow graphs). A domain-specific optimized CCA reached an average speedup of 2.21× for small application kernels [CBC+05].

In addition to these approaches, manifold coarse-grained reconfigurable approaches from academia (e.g., RAW [WTS+97], REMARC [MO99], PipeRench [GSB+00], HoneyComb [TB05]), and industry (e.g., Montium [HSM03, Rec] and PACT XPP [BEM+03, Tec06]) exist. However, this monograph presents a novel concept and architecture using a fine-grained reconfigurable fabric and thus, this chapter will focus on this type in the following. Before discussing general properties and presenting state-of-the-art reconfigurable processors, at first an overview of the underlying hardware architecture to realize fine-grained reconfigurable architectures is provided.

Figure 2.3 shows the internal structure of a Xilinx Virtex-II device [Xil07b] as an example for a fine-grained reconfigurable logic. The left side shows the so-called slice that comprises two 4:1 LUTs and two flip-flops together with multiplexers and connections. The LUTs can implement any functionality with four inputs and one

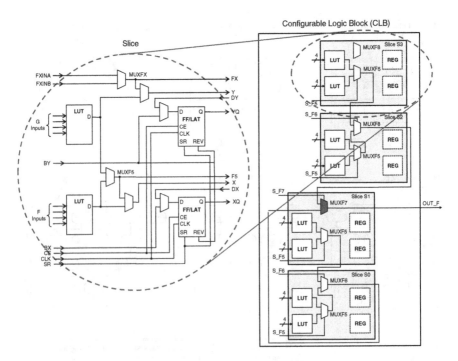

Fig. 2.3 Connecting LUTs as the basic building blocks of fine-grained reconfigurable logic to slices and configurable logic blocks

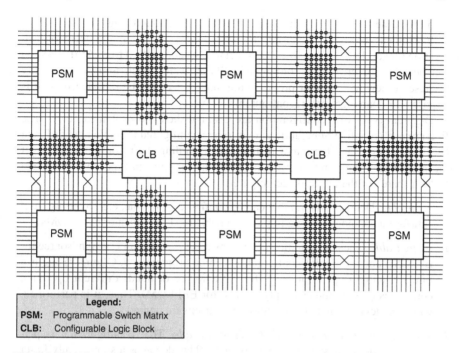

Fig. 2.4 Two-dimensional array of fine-grained reconfigurable CLBs that are connected with fine-grained reconfigurable switching matrices

output, i.e., all $2^4 = 16$ values of the corresponding truth table are configurable. To realize functions with more inputs or outputs, multiple LUTs need to be connected. The right side of the figure shows the so-called configurable logic block (CLB) that comprises four slices, i.e., eight LUTs. Here it becomes visible that some slices are directly connected with each other, which allows implementing a fast carry chain that connects multiple LUTs. Figure 2.4 shows the general structure of a fine-grained reconfigurable fabric that contains CLBs and programmable switching matrices (PSMs). The CLBs and PSMs are interconnected with dedicated wires. In addition, the PSMs are build of configurable interconnects, i.e., it is configurable which PSM input is connected to which PSM output.

These fine-grained reconfigurable fabrics provide a high flexibility, i.e., all hardware descriptions can be synthesized, placed, and routed for fine-grained reconfigurable fabrics (within their area restrictions). Bit- and byte-level computation and manipulation can be realized with rather few reconfigurable resources. In addition, small memories and finite state machines can be implemented using a fine-grained reconfigurable fabric. In addition, word-level computation can be realized; however, coarse-grained reconfigurable fabrics are typically more efficient for word-level computation as they are specifically optimized for them. Due to the high flexibility of fine-grained reconfigurable fabrics, all accelerators that might be used for an ASIP design can also be used for a fine-grained reconfigurable processor. Furthermore, the tool flow for designing accelerators for ASIPs and for a fine-grained reconfigurable fabric is the same, i.e., a hardware description language like VHDL or Verilog can be used. Therefore, the related work for automatic detection and creation for accelerators and so-called special instructions (SIs) from ASIP research can be used for creating reconfigurable SIs as well, which improves usability because no new tool flow or programming model needs to be introduced, as it is required for coarse-grained reconfigurable arrays.

Different types of reconfiguration are distinguished and explained below. The fine-grained reconfigurable fabric allows to be reconfigured to implement different hardware designs and the major distinctions according the reconfiguration target the questions "when and how often" a reconfiguration may be performed and "which share" of the fabric is reconfigured.

Configurable. Denotes a fabric that can be configured one time, e.g., by using fuse/anti-fuse technology as used in programmable read-only memories (PROMs).

Reconfigurable. The configuration can be erased (e.g., using an UV-light source for EPROM) and a different configuration can be programmed. Often, erasing is a time-consuming operation (e.g., for EPROM) or demands different operation conditions (e.g., a higher supply voltage for EEPROM). Thus, it is performed, when the device is not active, similar to a firmware upgrade.

Run-time reconfigurable/dynamically reconfigurable. The reconfiguration is performed while the system is up and running. This demands a significantly faster reconfiguration time (microseconds to milliseconds instead of minutes).

Partially run-time reconfigurable. Often, not the entire reconfigurable fabric needs to be reconfigured. Instead, only a certain accelerator shall be reconfigured into a region of it. At the same time, the parts that are not reconfigured remain functional and active.

In addition to this classification, also the question "who triggers the reconfiguration" leads to different concepts. However, this decision is often coupled with the above-described categories. For configurable and reconfigurable systems, typically the user has to initialize and observe the reconfiguration, whereas for (partially) run-time reconfigurable designs it is common that the design itself triggers the reconfiguration (using an on-chip available access to the configuration memory, e.g., the internal configuration access port (ICAP [Xil09c]) in case of Xilinx Virtex FPGAs). In the following, the discussion will focus on the class of partially run-time reconfigurable architectures. Nowadays, basically all architectures use the Xilinx FPGAs and tools [LBM+06] to prototype partial run-time reconfiguration in practice. Especially here, the reconfiguration time may become an issue because the reconfiguration is performed as part of the normal operation of the architecture and the application may have to wait until the reconfiguration is completed, as the reconfigured accelerator is demanded to proceed. The reconfiguration time depends on (a) the amount of configuration bits that need to be transferred and (b) the reconfiguration bandwidth. For fine-grained reconfigurable logic, typically tens to hundreds of kilobyte need to be transferred (depending on the size of the design and the size of the reconfigurable fabric that shall be reconfigured). To reconfigure an entire FPGA device, even megabytes need to be transferred. The reconfiguration bandwidth can be as slow as a few MB/s (depending on the performance of the memory that stores the reconfiguration data) or up to more than 100 MB/s [CMZS07]. The reconfiguration time is typically in the range of milliseconds. For instance, the extreme case of reconfiguring 10 KB at 100 MB/s (=100 KB/ms) demands 0.1 ms.[2] Vice versa, reconfiguring 100 KB at 1 MB/s demands 100 ms.[3] To improve the reconfiguration performance, bitstream compression was suggested to improve the effective transfer rate from off-chip memory to the on-chip configuration port [HMW04, HUWB04, KT04, LH01]. Under certain circumstances, it may even be possible to hide the reconfiguration time by starting it before it is needed. [LH02] analyzes concepts for configuration prefetching such that the reconfigurations are completed (in the best case) before the accelerators are required. They distinguish between static prefetching (the compiler determines "when" and "what" shall be reconfigure), dynamic prefetching (a run-time systems determines the decisions), and hybrid prefetching (a compiler-guided run-time system determines the decisions). However, the reconfigurable fabric is typically highly utilized, because typically all computational blocks exploit the parallelism of the reconfigurable fabric to accelerate their kernels. Therefore, insufficient free space might be available to start prefetching. In addition, the time between computational blocks is typically small compared to the reconfiguration time (otherwise, it would rather be another computational block that benefits from accelerators). Therefore, prefetching for a computational block after the previously executed computational block finished, does not hide the reconfiguration time.

2.2.2 Using and Partitioning the Reconfigurable Area

Fine-grained reconfigurable systems can be partitioned into general frameworks and specific processors. Typically, they differ in the complexity of accelerators that can be reconfigured, i.e., the general frameworks often envision entire IP cores or tasks that are provided as hardware implementation, whereas specific processors typically provide reconfigurable functional units that can be reconfigured to contain an application-specific accelerator (presented in the next sections), similar to those that are created and used by nonreconfigurable ASIPs.

An example for a general framework for reconfigurable IP cores/tasks [UHGB04b] is shown in Fig. 2.5. It partitions the reconfigurable fabric into reconfigurable modules and connects them using a bus. An arbiter controls the communication of the modules with each other and with the run-time system that provides access to external periphery (e.g., a CAN interface) and manages the reconfiguration of the modules. A different approach is the Erlangen Slot Machine (ESM, [MTAB07]). As shown in Fig. 2.6, it provides different types of intermodule communication, ranging from dedicated communication to direct neighbors and shared SRAM to a reconfigurable bus system [ESS+96]. All modules have a dedicated

---

[2] Please note that 10 KB is a rather small amount of configuration data and 100 MB/s is a rather fast reconfiguration bandwidth, i.e., this indicates a lower limit.

[3] The other extreme indicating an upper limit.

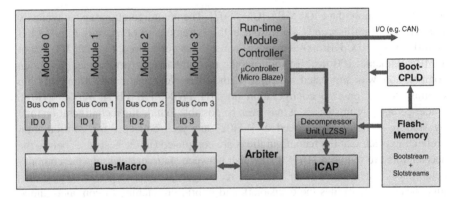

Fig. 2.5 Example for a general framework for partially reconfigurable modules that comprise dedicated IP cores or tasks

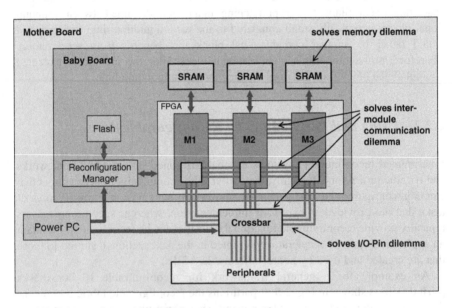

Fig. 2.6 Erlangen slot machine (ESM) architecture overview

connection to an off-chip crossbar that connects the modules to their external periphery without blocking the on-chip bus system. In addition, the reconfiguration manager and a PowerPC processor are available off-chip to execute software, start reconfigurations, and control the system.

It is noticeable that both approaches for general frameworks partition their reconfigurable fabric in modules (so-called partially reconfigurable modules, PRMs) and provide special interconnect structures for them. There are multiple technical and conceptual reasons for this composition.

Dedicated communication points. Run-time reconfigurable designs come with the conceptual problem that the synthesis and implementation tools (place and route) never see the entire design. This is because, in addition to the static parts of the design, some parts (the PRMs) are run-time reconfigurable and these parts are not necessarily known or fixed when the static part is synthesized and implemented. Therefore, the question arises how data can be exchanged between the static part and the PRMs. To solve this conceptual problem, dedicated communication interfaces (so-called bus macros) are placed at the borders between the static part and the PRMs. Then, the static design routes all communications to the bus macros, considering the content of the PRMs as a black box. Now, the designs for the PRMs can be synthesized and implemented independently, connecting their I/Os to the bus macros as well. The conceptual problem of establishing the communication limits the placement of the PRMs to those places, where bus macros are available and connected to the static part of the design.

Bitstream relocation. Considering that multiple PRMs are available and multiple IP cores/tasks shall be loaded into them, potentially multiple different partial bitstreams (i.e., the configuration bits to configure a PRM) need to be provided for each design that shall be loaded into any particular PRM. In the worst case, one bitstream needs to be provided for $task_i$ that shall be loaded into PRM_1, another bitstream to load $task_i$ into PRM_2, PRM_3, and so on. This potentially large amount of different bitstreams can be avoided when all PRMs have the same shape (i.e., outline and resource coverage). In this case, the concept of bitstream relocation can be used [BLC07, KLPR05]. Then, only one bitstream needs to be stored for each task and this bitstream can be adapted on the fly to be loaded into any particular PRM.

Writing the configuration data. To reconfigure a PRM, its configuration data needs to be changed. However, the configuration data is only accessible in so-called frames, i.e., a frame is reconfigured either entirely or not at all. For Virtex-II devices, a frame spans the whole height of the device and thus dictates that entire columns are reconfigured. That is also the reason why the PRMs in Figs. 2.5 and 2.6 correspond to horizontal slots. However, these technical constraints are specific to the particular FPGA family. For instance, in the Virtex-4 and the Virtex-5 FPGAs, the size of a frame is changed and it spans a height of 16 and 20 CLBs, respectively.[4] Independent of the height of the frame, a PRM should always be adapted to it, i.e., it should have the same height. It is possible to reconfigure only parts of a frame by first reading the configuration data of the frame, then modifying the fraction that shall be reconfigured, and writing back the entire frame [HSKB06]. However, that increases the delay of the reconfiguration process significantly (for Virtex-II devices, the obtained flexibility may overcome the overhead). Therefore, this technical constraint also limits the shape of PRMs, namely to rectangular outlines that are aligned at frame and CLB borders.

Summarizing, the conceptual demand for communication points, the support for bitstream relocation, and the limited access to the reconfiguration data favor certain

---

[4] Multiple frames need to be reconfigured for a CLB column, i.e., the width of a frame does not cover a CLB.

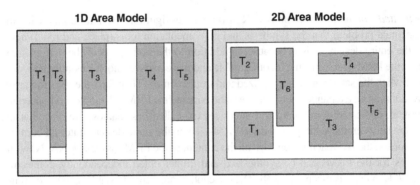

Fig. 2.7 Different area models for fine-grained reconfigurable fabric

types of designs, i.e., rectangular PRMs that are aligned to configuration frames and CLBs that have identical sizes and outlines and that provide dedicated communication points.

In addition to the full-height reconfigurable PRMs (i.e., a 1D placement of the PRMs), also a 2D placement [SWP04] is possible, when using a Virtex-4/5 FPGA or applying the read–modify–write technique from [HSKB06], as for instance demonstrated in [SKHB08]. Figure 2.7 illustrates the conceptual differences. The 2D area model is especially beneficial, if the designs that shall run on the reconfigurable fabric correspond to entire tasks (e.g., managed in [LP07]) and their hardware implementations are rather different in size. In such a scenario, it cannot be predetermined "where" a particular functionality will be placed on the FPGA. A run-time placement of tasks to an FPGA is presented in [WSP03]. A high fragmentation can lead to the undesirable situation that a task cannot be placed although there would be sufficient area available. Therefore, a service that organizes the placement of tasks (in order to avoid such situations) is investigated. The goals of a fast task placement with a high placement quality are achieved by the management of the free rectangles with a hash matrix. The hash matrix allows finding a suitable rectangle in constant time, at the cost of a subsequent updating phase of this data structure. A routing-conscious dynamic placement for reconfigurable devices is presented in [ABF+07] to reduce the communication latency. However, the actual implementation of the communication infrastructure is not addressed, i.e., how the static part can be connected to the task that is loaded into the dynamic part if this task may be placed anywhere in the dynamic part, irrespective of dedicated communication points.

One possibility to establish the communication in a 2D area model is using a network-on-chip (NoC) architecture that provides dedicated communication points at various places on the reconfigurable fabric, as proposed by DyNoC [BAM+05] shown in Fig. 2.8. When a PRM is reconfigured and it covers these communication points, then the corresponding router is disabled and an adaptive routing strategy is

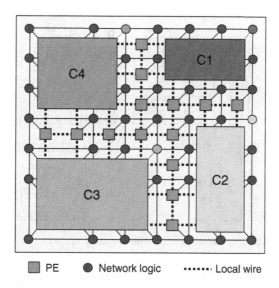

Fig. 2.8 2D area model using a dynamic network-on-chip architecture to establish communication

used to assure that the data packets still arrive at their destination.[5] CoNoChi [PKAM09] proposes a NoC architecture that allows reconfiguring the structure of the NoC, i.e., it provides a concept and an implementation approach that allow adding or removing routers during run time. Instead of disabling routers, it allows to remove routers and change the connection topology of the network. For instance, on the right side of PRM C2 in Fig. 2.8, altogether four routers are placed. That means that each data packet that moves along that path demands four hops to traverse that distance. The concept of CoNoChi allows removing the intermediate routers and establishing a direct connection between the routers at the borders to improve the packet latency.

2.2.3 Coupling Accelerators and the Processor

As discussed in the previous section, the noticeable overhead of a flexible 2D area model is mainly beneficial if the tasks that shall be reconfigured differ in size significantly. NoCs can be used to connect the different PRMs. However, when targeting application-specific reconfigurable accelerators (similar to ASIPs), then a NoC is not necessarily the best choice to connect the accelerators with the core processor as the NoC introduces a nonlegible communication latency. Here, different alternatives exist to couple the core processor with the reconfigurable accelerator. Figure 2.9 shows conceptually different examples that vary in the amount of required processor modification and provided latency and bandwidth [CH02, TCW+05].

---

[5]This requires that around each PRM a ring of routers remains active.

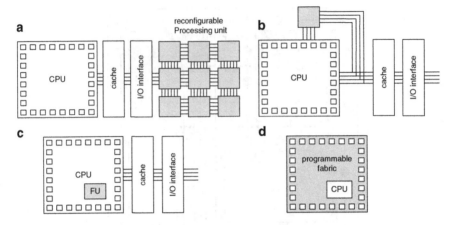

Fig. 2.9 Coupling reconfigurable accelerators with the core processor

Figure 2.9a shows a rather loose coupling, where the reconfigurable fabric is off-chip, attached via an I/O interface. The advantage is that it can be coupled to any existing chip that provides access to an I/O interface; however, the communication bandwidth and latency is limited. Therefore, a performance improvement can only be expected if a rather large amount of computation needs to be performed on a small amount of data, thus diminishing the effects of the limited communication performance. Figure 2.9b shows an on-chip connection between the reconfigurable fabric and a core processor. This connection may use a typical coprocessor interface for connection. The PRM may also be connected after the cache, depending on the provided interface of the targeted core processor. This also corresponds to the type of connection that a NoC communication infrastructure would use. The tightest coupling is shown in Fig. 2.9c. The reconfigurable fabric is embedded into the pipeline of the core processor as a reconfigurable functional unit. To establish this connection, it is required to change the processor-internal structure. The advantage is the communication bandwidth and latency, because direct connection to the register file is possible, i.e., multiple words can be read and written per cycle. This corresponds to the coupling that is used for ASIPs and typically, the amount of read/write ports of the register file is increased to extend the bandwidth even further. Figure 2.9d shows a combination that is often used for prototyping but may also be used as dedicated target architecture. An FPGA device is used as reconfigurable fabric and the core processor is implemented as nonreconfigurable part within the reconfigurable fabric (either as "hard" core or using the reconfigurable fabric).

For the architecture that is presented in this monograph, the tight coupling of Fig. 2.9c is targeted; however, for prototyping purpose the type of Fig. 2.9d is used. This tight coupling corresponds to the basic idea that reconfigurable accelerators shall be provided in a similar way like ASIPs use them. Therefore, the advantage of a rather simple area model with fixed communication points and similar-sized PRMs outpaces the rather complex 2D area model with its NoC communication paradigm. Still, a 2D placement of PRMs is used, but the communication points and PRM placements are fixed.

2.2.4   Reconfigurable Instruction Set Processors

In the scope of the work that is presented in this monograph, the focus is placed on extensions of CPUs with reconfigurable functional units (RFUs) that are tightly coupled to the core pipeline. An overview and classification for this specific area of reconfigurable computing is given in [BL00]. In this section, state-of-the-art approaches that are related to the reconfigurable processor that is presented in this monograph are presented and discussed.

Employing reconfigurable processors affects the application development in two different ways. First, the application-specific hardware accelerators need to be designed for the RFUs (typically in a hardware description language) and afterwards, these accelerators need to be operated efficiently. To obtain a benefit from the hardware accelerators, the application programmer has to insert special instructions (SIs) into the application (e.g., using inline assembly or compiler tools) as a way to access the accelerators.[6] In a way, both steps are comparable to the development process of ASIPs, which allows that their concepts for tool flows and automatic development environments may be adapted for this class of reconfigurable processors. However, reconfigurable processors additionally need to determine which accelerators should be loaded into the RFU at which time during the application execution. This is often accomplished by so-called prefetching instructions [LH02] that trigger the upcoming reconfigurations.

The difference between ASIPs and reconfigurable processors becomes noticeable when executing SIs. An ASIP provides all SIs statically, typically using the same fabrication technology like the core pipeline. A reconfigurable processor instead uses a fine-grained reconfigurable fabric for the SIs. In addition, a reconfigurable processor may not have an SI available when it shall execute, e.g., because the process of reconfiguration has not finished yet. The partitioning of the reconfigurable fabric (e.g. how many SIs can be provided at the same time) has a high impact on whether or not an SI is available when it is demanded. Furthermore, architectural parameters (for instance the reconfiguration time) affect the ability to efficiently provide SIs on demand. All these aspects are common for the following state-of-the-art architectures (presented in chronological order); however, they address them using different techniques.

The *OneChip* project [WC96] and its successor OneChip98 [CC01, JC99] use RFUs within the core pipeline to utilize reconfigurable computing in a CPU. The RFUs are accessed by SIs as multicycle instructions. As their speedup is mainly obtained from streaming applications [CC01] they allow their RFUs to access the main memory, while the core pipeline continues executing. OneChip is based on an in-order

---

[6] Note that further access possibilities – e.g., memory-mapped interfaces or co-processor ports – may be used as well; however, for tightly coupled accelerators, the SIs provide a direct and low-latency interface as it is used by ASIPs.

pipeline, whereas OneChip98 uses an out-of-order superscalar architecture. Therefore, memory inconsistency problems may occur when an SI that accesses the memory executes on the RFU and – during the multicycle SI execution – load/store instructions execute in the core pipeline. For instance, the pipeline may load a value from memory, modify it, and write it back. If an SI writes the same address in between, then the core pipeline overwrites its result. In the scope of the OneChip98 project, nine different memory inconsistency problems were identified and hardware support was developed to resolve them automatically. The RFUs provide six configuration contexts and can switch between them rather fast. To load an RFU configuration into one of these contexts, the access to the main memory is used.

The *Chimaera* project [HFHK97, YMHB00] couples a dynamically scheduled superscalar processor core with a reconfigurable fabric that may contain multiple SIs. The SIs obtain up to nine inputs and create one output value directly from/to the register file (i.e., Chimaera uses a tight coupling). To avoid implementing a register file with nine read ports, these nine registers are realized as a partial physical copy of the register file that is updated when the corresponding registers in the register file are modified (so-called shadow registers). The register addresses are hard coded within the SI and the compiler/application programmer is responsible for placing the expected data in the corresponding registers. Chimaera uses a customized reconfigurable array that is based upon 4:1 LUTs that can also be used as two 3:1 LUTs or one 3:1 LUT with additional carry computation. A fast carry logic is shared among all logic blocks in a row and the routing structure of the array is optimized for arithmetic operations. When an SI demands the reconfigurable array but the required configuration is not available, then a trap is issued and the trap-handler performs the reconfiguration. Therefore, the application computation is stalled during the array reconfiguration and in the case of large working sets (i.e., many SIs within a loop), the problem of thrashing in the configuration array is reported (i.e., frequent reconfigurations within each loop iteration). Therefore, Chimaera provides a dedicated on-chip configuration cache to diminish the effects of this problem.

The *CoMPARE* processor [SGS98] provides a reconfigurable fabric that is coupled to the pipeline of the core processor similar to an ALU. The reconfigurable fabric implements an RFU that takes four inputs and returns two results. The RFU does not provide any registers or latches, i.e., it is stateless. Therefore, no state-machine or data feedback may be implemented. In addition, the RFU can only comprise the hardware implementation of a single SI at a time. If more SIs are demanded then the RFU needs to be reconfigured in between their executions.

The *Proteus* processor [Dal99, Dal03] extends a processor by a tightly coupled fine-grained reconfigurable array that is divided into multiple RFUs, where each RFU may contain one SI implementation at a time. Proteus concentrates on operating system support with respect to SI opcode management. This is especially beneficial when multiple tasks need the same SI, as they can share the same SI implementations without the need that the tasks were compiled such that they use the same opcode for the SI. However, when multiple tasks exhibit dissimilar processing characteristics, a task may not obtain a sufficient number of RFUs to execute all SIs in hardware. Therefore, some SIs will

execute in software, resulting in steep performance degradation. In addition to SI management, Proteus provides concepts for task switches that may occur during an SI
execution, i.e., the context of the SI that currently executes needs to be stored, saved,
and restored (when the task and thus the SI continues execution later on).

The *XiRisc* project [LTC+03, MBTC06] couples a VLIW processor with a
reconfigurable gate array. The instruction set architecture of the VLIW processor is
extended to provide two further instructions for reconfiguring the array and for
execution SIs. The SIs have access to the register file and the main memory and
may last multiple cycles. The configuration of an SI is selected out of four different
contexts and reconfiguration between them can be done in a single cycle. These
multiple contexts are beneficial if small applications fit into them. In [LTC+03] the
fastest reported speedup (13.5×) is achieved for DES and the only context reloading
happened when the application was started. However, in [MBTC06] a relevant
MPEG-2 encoder is used for benchmarking. Here, run-time reconfiguration is
required (as the accelerators no longer fit into the available contexts) and the
achieved speedup reduced to 5× compared to the corresponding processor without
reconfigurable hardware (i.e., GPP).

The *Molen* processor [PBV06, PBV07, VWG+04] couples a reconfigurable fabric
to a core processor via a dual port register file and an arbiter for shared memory. The
core processor is extended to provide additional instructions to manage the reconfigurations and SI executions. The instruction set extension can be parameterized to support only one or multiple SIs at a time, i.e., they provide instructions that allow to
perform partial reconfiguration, but not all Molen instances necessarily use them (i.e.,
they may restrict to instructions that reconfigure the entire reconfigurable fabric). The
run-time reconfiguration is explicitly predetermined by these control instructions that
also allow configuration prefetching. The compiler/application developer is responsible for deciding which configuration shall be loaded at which time. All SIs are
executed by a generic instruction that obtains the address of the configuration bitstream as parameter to identify the demanded SI. Therefore, the amount of SIs is not
limited by the number of available opcodes of any particular instruction set. To provide parameters to the SIs, data has to be explicitly moved to dedicated exchange
registers that couple the reconfigurable fabric to the core pipeline.

The *Warp* processor [LSV06, LV04] automatically detects kernels (using online
monitoring) while the application executes. Then, custom logic for these kernels is
generated at run time through on-chip micro-CAD tools (i.e., synthesis, place, and
route). These kernel implementations are then called from the application. However,
as the application does not explicitly contain SIs in the binary (as the kernel was not
determined at compile time yet), Warp proposes a different approach. They replace
parts of the application binary (that correspond to the kernel that shall execute using
the reconfigurable fabric) by a hardware initialization code. This code provides input
data to the kernel implementation, starts its execution, and stalls the processor. After
the kernel execution finished, the processor receives an interrupt and continues
where it stopped, i.e., in the hardware initialization code. Directly following that
code, a jump instruction is placed to continue the application execution with the code

that follows the kernel. Warp proposes a specialized fine-grained reconfigurable fabric that trades the performance of the reconfigurable fabric with the computational complexity for the on-chip micro-CAD tools.[7] They focus on the memory requirements of the CAD tools, as this is a limited resource in embedded systems. Still, the online synthesis incurs a nonnegligible computation time and therefore the authors concentrate on scenarios where one application is executing for a rather long time without significant variation of the execution pattern. In these scenarios, only one online synthesis is required (i.e., when the application starts executing) and thus the initial performance degradation amortizes over time.

2.3   Summary of Related Work

Application-specific instruction set processors (ASIPs) allow accelerating applications by providing dedicated hardware accelerators. Their execution is triggered by special instructions (SIs). Research in the scope of fine-grained reconfigurable processors focused on connecting a core processor with an FPGA-like reconfigurable fabric on which SI implementations are dynamically loaded during run time. The presented state-of-the-art approaches mainly concentrated on offering and interfacing SIs in their reconfigurable fabric. Some of them only provide one SI at a time and some support multiple SIs that are available on the reconfigurable fabric at the same time. Still, all of the above-discussed approaches potentially increase the utilization of the available hardware resources by reconfiguring parts of it to match the current requirements of the application (i.e., the currently demanded SIs). However, due to the reconfiguration time the utilization of the reconfigurable fabric may often be suboptimal.

All presented approaches only consider one predetermined implementation per SI. The Proteus reconfigurable processor additionally offers the execution of an SI with the cISA, but the operating system predetermines which SI shall be executed with reconfigurable hardware and which with the cISA. Still, there is at most one hardware implementation per SI. The Warp processor could potentially change the SI hardware implementations by performing further online syntheses. However, due to the implied overhead, synthesizing SI implementations is typically only done once after the application started, which finally leads to a single SI implementation as well.

In addition, state-of-the-art fine-grained reconfigurable processors mainly focus on compile-time predefined reconfiguration decisions. Often, dedicated commands have to be inserted into the application binary by the compiler (or the application programmer) that explicitly determine "when" the reconfiguration of the hardware implementation of "which" SI shall be started. This is not suitable when computational requirements or constraints are unpredictable during compile time and change during run time.

---

[7] A more regular reconfigurable fabric with limited interconnect possibilities simplifies place and route at the cost of reduced SI performance.

These and similar concerns are addressed by the paradigm of the rotating instruction set processing platform (RISPP) that is presented in this monograph and that fixes some customizations during compile time and is able to adapt dynamically during run time. Therefore, this monograph presents the novel concept of modular SIs that allows providing multiple hardware implementations per SI. A concept plus an infrastructure for modular SI is developed that provides the feature to upgrade dynamically from one implementation of an SI to another. This alleviates the problem of long reconfiguration time whenever prefetching is infeasible and additionally offers high flexibility to adapt to different requirements (e.g., by deciding which SI shall be upgraded further) that depend on the run-time situation. Furthermore, this monograph presents a novel run-time system that uses the modular SIs to enable an adaptive and efficient utilization of the available hardware without statically predetermined reconfiguration decisions. The detailed problems of state-of-the-art and the concept and basics of modular SIs are explained in Chap. 3, whereas the algorithms of the novel run-time system are presented in Chap. 4. Chapter 5 presents the hardware architecture and its implementation whereas Chap. 6 provides overall benchmarks and comparison to state-of-the-art approaches, i.e. ASIPs and reconfigurable processors.

Chapter 3
Modular Special Instructions

This chapter presents the novel concept of modular special instructions (SIs) that provides the potential for improved adaptivity and efficiency in comparison to state-of-the-art monolithic SIs. The concept of modular SIs represents the foundation of the work presented in this monograph and the other chapters build upon this foundation. Their potential is utilized by the novel run-time system that is described in Chap. 4 and they are implemented using a novel computation and communication infrastructure that is described in Chap. 5.

The first section will analyze state-of-the-art monolithic SIs, identify multiple relevant problems that limit the performance and adaptivity of reconfigurable processors and discuss them. The next section presents the concept of modular SIs and it is shown how they address the beforehand-identified problems and what additional advantages they provide. The third section demonstrates multiple real-world examples of modular SIs for an H.264 video encoder that will also be used to evaluate the proposed RISPP architecture and to compare it with state-of-the-art architectures in Chap. 6. The last section presents a formal model that allows describing modular SIs in a precise and clear way. This model is used to describe the tasks of the run-time system and the pseudo codes of the proposed algorithms in Chap. 4.

3.1 Problems of State-of-the-Art Monolithic Special Instructions

A special instruction (SI) is an assembly instruction that implements an application-specific functionality to accelerate the execution of a particular application or an application domain (see also Sect. 2.1). The complexity of SIs reaches from rather small instructions (e.g., *multiply and accumulate*) to rather large ones (e.g., *discrete cosine transformations*, as required for many image and video compression standards). The implementation of an SI is not restricted to the computational hardware that is available in the processor pipeline. Instead, typically an extra hardware block is designed and added to the processor. The assembler instruction provides an interface between the application and this hardware block.

L. Bauer and J. Henkel, *Run-time Adaptation for Reconfigurable Embedded Processors*,
DOI 10.1007/978-1-4419-7412-9_3, © Springer Science+Business Media, LLC 2011

In state-of-the-art reconfigurable processors (e.g., Molen [VWG+04], Proteus [Dal03], OneChip [CC01], etc.) an SI corresponds to a hardware block that is implemented using a reconfigurable fabric. To be able to reconfigure SI implementations at run time, the reconfigurable fabric is partitioned into reconfigurable regions, so-called reconfigurable functional units (RFUs), that are connected to the processor pipeline. To be able to load any SI implementation into any RFU, the interface between the RFUs and the processor pipeline needs to be identical for all RFUs. Please note that introducing multiple different RFU types with different interfaces does not generally solve this limitation, even though it reduces its effects. However, introducing different RFU types would create a fragmentation problem, e.g., it may not be possible to load an SI into the reconfigurable fabric even though an RFU is free because the interfaces of SI and RFU do not match.

Nonreconfigurable extensible processors define during design time which SIs the processor shall support. As this well-defined set of SIs is fixed after the processor is taped out, it is possible to apply global optimizations when designing the SI implementation, i.e., a combined hardware module that implements all SIs can be created. This approach allows sharing of common computational parts of the SIs, i.e., sharing of data paths (e.g., adder, multiplier, etc.). For instance, when multiple SIs demand an adder, then not necessarily multiple adders are needed for their implementation when these SIs never execute in parallel. However, data path sharing is not always possible for processors with a reconfigurable instruction set. This sharing problem is due to concepts and properties of run-time reconfiguration (but alternatives exist, as shown afterwards):

- On the one hand, it is not necessarily known which SI implementations will be available (i.e., loaded to the reconfigurable fabric) at the same time during application execution. This highly depends on the decisions which SI implementations shall be reconfigured at which time. Therefore, it is not known, which SI implementations might share a common data path (both implementations need to be loaded at the same time) and which need a dedicated implementation of a potentially common data path.
- On the other hand, reconfigurable SI implementations demand a fixed implementation interface to be able to load any SI implementation into any RFU (as described above). Therefore, to support data path sharing, each possibility to share a data path demands an extension of the general RFU interface which complicates its design. Vice versa, for a predetermined RFU interface the sharing potential is limited.

Due to these reasons, all state-of-the-art reconfigurable processors that target reconfigurable SIs, target so-called monolithic SIs, i.e., each SI is implemented as a dedicated hardware block (no support for sharing) that is loaded to the reconfigurable fabric either entirely or not at all. During the reconfiguration process – that can last multiple milliseconds (e.g., in [GVPR04]) – the SI implementation that was previously loaded into the RFU is no longer usable (parts of it are overwritten) and the SI implementation that is currently being reconfigured is not usable yet (it is not completely reconfigured yet).

In addition to the potentially increased hardware requirements (as sharing is not supported, redundant instances of the same data path might be needed), the concept of monolithic SIs – as used by all state-of-the-art processors with support for reconfigurable SIs – has two further drawbacks that are due to the concept of monolithic SIs. To allow each SI to be loaded into any RFU (as described above) all RFUs need to provide the identical amount of reconfigurable fabric. If the hardware implementation of an SI is relatively small in comparison to an RFU this leads to a noticeable *fragmentation problem*, because the share of the RFU that is not required by its loaded SI cannot be used to implement another SI. Similar to the sharing problem, the fragmentation problem potentially leads to increased area requirements (corresponding to decreased area efficiency).

However, not only the area requirements are affected, also the performance is potentially affected in a negative way. The noticeably increased performance of (reconfigurable) SIs is based on the increased parallelism of their implementations. More parallelism (limited by area constraints and input data bandwidth) leads to a higher performance. However, more parallelism also demands more hardware and a correspondingly larger SI implementation results in a longer reconfiguration time. Therefore, depending on the reconfiguration frequency,[1] a highly parallel implementation might lead to a reduced performance in comparison to an SI implementation with limited parallelism, because the *reconfiguration overhead problem* diminishes the potential performance to some degree. For instance, if the software implementation of a computational block demands 10 ms execution time (ET) and a dedicated SI provides a 10× speedup for this computation but demands 4 ms reconfiguration overhead (RO) until the computation can start, then the resulting speedup is *10 ms ET/(4 ms RO + 1 ms ET) = 2×*. Considering an SI implementation that leads to 5× speedup but demands only 2 ms reconfiguration overhead, the resulting speedup improves to *10 ms ET/(2 ms RO + 2 ms ET) = 2.5×*. This shows that the beneficial amount of parallelism is limited by the reconfiguration overhead.

Reconfigurable processors typically address the reconfiguration overhead problem by supporting the following two techniques:

1. *Prefetching SI implementations.* Prefetching aims to start the reconfigurations of a demanded SI implementation as early as possible. The decision which reconfiguration is started and when it is started is determined at compile time (static prefetching), at run time (dynamic prefetching), or at run time using compile-time-provided information (hybrid prefetching) [LH02].
2. *cISA execution of SIs.* Whenever an SI is required to execute but the SI implementation is not available (e.g., because the reconfiguration is not completed yet), the SI may be executed using the core Instruction Set Architecture (cISA) of the pipeline. This means that the statically available hardware (e.g., the ALU

---

[1] If an implementation is reconfigured only once when the application starts, then the reconfiguration overhead will amortize over time; however, if frequent reconfigurations are demanded, it might not amortize.

in the pipeline) is used by executing instructions from the cISA, similar to the execution on a general-purpose processor without SIs (see also Sect. 5.2).

Even though these two techniques potentially reduce the negative effects of the reconfiguration overhead problem, they are not always effective, as shown later. The cISA execution avoids stalling the pipeline unnecessarily long.[2] However, it executes the functionality of the SIs rather slow (at most as fast as without using SIs) and thus does not lead to a noticeable performance improvement. In addition, prefetching is only possible when the following two conditions apply:

1. The upcoming SI requirements are known with a good certainty and early enough (multiple milliseconds before the actual SI execution).
2. A sufficiently large reconfigurable fabric is available such that – in addition to the at-that-time required SIs – some RFUs are available to prefetch the implementation for the upcoming SI requirements. Otherwise, some of the still-required SIs need to be replaced which would affect the current SI requirements and thus makes prefetching unbeneficial.

In Fig. 3.1, different trade-offs between SI parallelism and reconfiguration overhead are compared, showing the SI execution pattern for the motion estimation (ME) of an H.264 video encoder (see also Sect. 3.3). The x-axis shows a time axis (the ME execution starts at time 0) and the y-axis shows how many SIs finished execution since time 0. Altogether, 31,977 executions of two different SIs are demanded for that particular motion estimation (depending on the motion estimation algorithm and the input video). Line ① shows the SI execution pattern

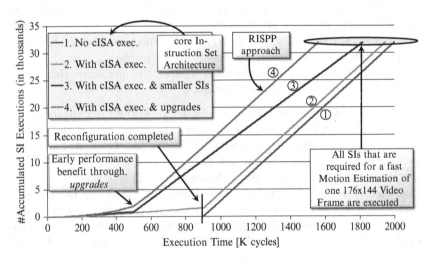

Fig. 3.1 Comparing different performance vs. reconfiguration overhead trade-offs

---

[2]Without cISA execution, the pipeline would need to be stalled until the reconfigurations that are demanded to execute the SI completed.

for a reconfigurable processor that does not support cISA execution and that uses an SI implementation that finishes reconfiguration 900,000 cycles after the computational hot spot started execution (corresponds to 9 ms for a 100 MHz clock frequency). Therefore, in the first 900,000 cycles, the pipeline stalls until the reconfiguration completes. Line ② introduces the concept of a cISA execution, leading to a small performance improvement. Instead of stalling the pipeline, the execution can proceed, even though at a noticeably smaller performance in comparison to the later hardware execution (visible by the gradients of line ② before and after the reconfigurations are completed). Line ③ shows the effect of a different SI implementation that exploits less parallelism and therefore finishes the demanded reconfiguration earlier. After 500,000 cycles the reconfiguration completes and the SI implementation switches from the cISA execution to a parallel hardware implementation. As long as line ② executes the SIs in cISA, the performance gain of line ③ in comparison to line ② increases. However, as soon as line ③ switches from cISA execution to hardware execution, this initial performance gain decreases (due to a less parallel SI implementation of line ③). In this particular example, line ③ leads to the faster execution. However, in Fig. 3.1 it becomes visible that this highly depends on the total amount of SI executions and – for a motion estimation algorithm – this typically depends on the input video data. Eventually, line ④ gives an outlook on the SI execution pattern for the rotating instruction set processing platform (RISPP) approach that is explained in the following section. It combines the performance improvement of line ③ during the early stage of the execution (and even outperforms it) with the parallel implementation of line ① and ② in the later stage of the execution by the concept of *upgrading* the SI implementations (details in Sect. 3.2).

Altogether, this monograph identified and described the following three different problems for state-of-the-art reconfigurable processors with monolithic SIs:

1. *Sharing problem.* Sharing of common data paths is only possible under constraints that would significantly limit the adaptivity. Thus, state-of-the-art reconfigurable processors do not support sharing.
2. *Fragmentation problem.* To be able to reconfigure SI implementations at run time in a flexible way and to be able to connect them to the processor pipeline, a certain area budget and connection interface has to be used for each RFU. If an SI implementation is relatively small in comparison to the provided reconfigurable fabric within the RFU, the remaining resources cannot be used to implement another SI.
3. *Reconfiguration overhead problem.* The reconfiguration overhead depends on the amount of parallelism that is exploited in the SI implementation. Therefore, the amount of effectively usable parallelism is limited. In addition, this effectively usable amount depends on the expected number of SI executions.

It is noticeable that none of these three problems applies to state-of-the-art extensible processors (see Sect. 2.1). They provide all SIs in a nonreconfigurable implementation. Therefore, each SI may use an individual interfaces to the processor pipeline and the SIs may share common data paths without affecting these interfaces.

In addition, no spare area has to be reserved to load potentially larger SI implementation. Therefore, no fragmentation occurs. Eventually, as all SI implementations are statically available, no reconfiguration overhead reduces the amount of usable parallelism. However, reconfigurable processors also have noticeable advantages: they are no longer fixed to a certain set of SIs. Instead, an application developer can create new SIs (or new implementations of an SI) that can be loaded onto the reconfigurable fabric at run time to accelerate further applications or application domains that were not necessarily considered when designing the reconfigurable processor. In addition to the adaptivity to support new SIs, reconfigurable processors provide run-time adaptivity. Depending on the application requirements (e.g., SI execution frequencies) the set of currently available SIs may be dynamically changed by reconfiguring those SIs that provide the highest benefit at a certain time during application execution.

Therefore, the aim is to combine the advantages of state-of-the-art extensible processors and state-of-the-art reconfigurable processors and even improve upon these. The approach to implement SIs is different in comparison to reconfigurable processors. In Sect. 3.2, this monograph will introduce *modular* SIs (in contrast to state-of-the-art *monolithic* SIs) that are based on a hierarchical SI composition. They provide a higher adaptivity than state-of-the-art monolithic reconfigurable SIs and they significantly reduce the three above-discussed problems (sharing, fragmentation, and overhead) that were discussed in this section.

3.2 Hierarchical Special Instruction Composition

State-of-the-art processors with a reconfigurable instruction set use so-called *monolithic* special instructions (SIs) as described in Sect. 2.2. In the concept of *monolithic* SIs, each SI is realized by one hardware implementation that can be loaded into an reconfigurable functional unit (RFU) during application run time. This single SI implementation corresponds to a certain trade-off between parallelism and size and thus determines the minimal size of an RFU (to be able to load that SI into it) and reconfiguration overhead.

Instead of providing a single implementation per SI, the presented approach proposes a concept that provides multiple implementations of an SI that differ in their parallelism/overhead trade-off but that all realize the same functionality. This concept additionally provides an extended level of adaptivity, as it allows switching from one SI implementation to another during run time, depending on the application requirements (e.g., how often the SI is demanded). Furthermore, each SI implementation is not a monolithic block, but it is a modular composition out of elementary data paths.

Figure 3.2 shows an abstract example of the proposed hierarchical SI composition, consisting of three levels, namely SIs, Molecules, and Atoms:

Atom. An atom corresponds to an elementary data path. It is the smallest reconfigurable unit, i.e., an atom is either completely reconfigured onto the reconfigurable

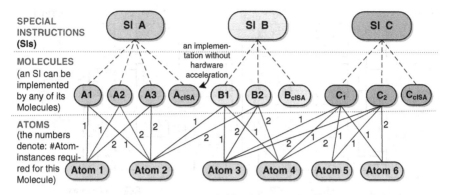

Fig. 3.2 Hierarchical composition of special instructions: multiple implementation alternatives – so-called molecules – exist per special instruction and demand atoms for realization

fabric or not (that is the reason for its name, i.e., it is considered as indivisible). The reconfigurable fabric is partitioned into so-called *atom containers* (ACs) and each AC can be dynamically reconfigured to comprise any particular atom (similar to RFUs but smaller in size, see Sect. 3.1). Each AC provides a fixed interface, comprising two 32-bit inputs, two 32-bit outputs, a clock signal, and a 6-bit control signal that can be used for atom-internal configurations, as shown later. This particular interface is not an integral part of the general concept, but it performed well in practice (examples are given below and in Sect. 3.3). In the following, atom type and atom instance are distinguished. The *atom type* denotes the functionality of the atom. The *atom instance* corresponds to an instantiation of an atom type. This means that multiple atom instances of a certain atom type can be available at the same time (i.e., reconfigured into different ACs).

Molecule. A molecule is a combination of multiple atoms and corresponds to an implementation of a special instruction. Typically, a molecule is composed of multiple different atoms types and each atom type may be demanded in different quantities. Therefore, a molecule corresponds to one specific trade-off between parallelism and reconfiguration overhead. The more atoms it demands, the more computations may be performed in parallel (as is shown later), but also more reconfigurations need to be performed. A molecule is available (i.e., completely loaded onto the reconfigurable fabric) if all of its demanded atoms are available. However, an atom is not dedicated to any molecule. Instead, a particular atom can be used to implement different molecules (independent whether or not they belong to the same SI as shown in Fig. 3.2) if these molecules are not executing at the same time.

Special instruction. A special instruction has the same semantic like in extensible or reconfigurable processors. It corresponds to an assembly instruction that accelerates the application execution. Modular SIs are available in multiple different implementations (i.e., molecules). Therefore, the implementation of a particular SI can be *upgraded/downgraded* during run time by switching from one molecule to another. Essentially, each SI corresponds to an acyclic graph where each node represents the functionality of an atom. A molecule determines a resource constraint

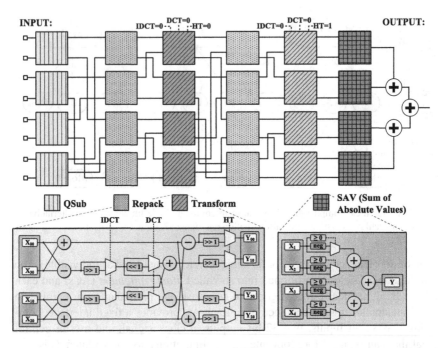

Fig. 3.3 Example for the modular special instruction SATD (sum of absolute (Hadamard-) transformed differences), showing the details for the Transform atom and the SAV (sum of absolute values) atom

for the graph, i.e., which atom types are available in which quantities. This affects the implementation of the SI[3] and thus it determines the latency (i.e., the demanded number of cycles) of the SI execution. In addition to the different hardware molecules, each SI also contains the so-called cISA molecule that does not demand any accelerating atoms but executes the SI functionality using the cISA of the processor pipeline (as described in Sect. 3.1).

Figure 3.3 shows an example SI as a graph of connected atoms. It calculates the sum of absolute Hadamard-transformed differences (SATD), as it is required in the motion estimation of an H.264 video encoder (more examples are presented in Sect. 3.3). On the left side in Fig. 3.3, the input data is provided to the SI (in this example the input data is loaded from data memory) and on the right side, the result is returned. The atoms (nodes of the graph) are connected together to form a combined data path that calculates the result of the SI. The atom-internal data path of the Transform atom and the sum of absolute values (SAV) atom are shown below the graph. It is noticeable that some operations within an atom work in parallel (e.g., the adders and subtractors at the input of the Transform atom). Other operations work sequentially within the same cycle (an atom typically completes operation

---

[3]This corresponds to a scheduling problem from the domain of high-level synthesis [Tei97]: the data-flow graph that describes the SI functionality is scheduled considering the resource constraints.

within one cycle; however, multicycle atoms are supported as well), using operation chaining. This is called atom-level parallelism, as a single atom already provides some degree of parallelism to accelerate the SI execution.

In addition to the atom-level parallelism, modular SIs exploit the so-called molecule-level parallelism. The smallest hardware molecule (regarding the number of atom instances) utilizes one instance per atom type that is used in the SI graph. The *smallest* hardware molecule for SATD in Fig. 3.3 demands one instance of QSub, Repack, Transform, and SAV, respectively. These four atoms are sufficient to realize the functionality of SATD in hardware. In this molecule, the instance of the Transform atom is used eight times to realize the eight different transform nodes in the SI graph. If two instances of the Transform atom would be available (corresponding to a *larger* molecule), then they may be used in parallel (i.e., higher molecule-level parallelism) to expedite the SI execution.

The molecule-level parallelism is an important feature of modular SIs, because it allows efficient SI *upgrading*. For instance, when one instance is available for all demanded atom types (*smallest* hardware molecule), loading an additional atom may lead to a faster molecule. Loading another atom might improve the performance further, and so on. This means, instead of loading these faster molecules from scratch (i.e., not considering the already available atoms) only one additional atom needs to be loaded, which makes the process of upgrading efficient. Table 3.1 provides an overview of different SATD molecules. In addition to the cISA molecule (319 cycles per execution), mixed molecules (explained in next paragraph) and hardware molecules exist. The smallest hardware molecules requires 22 cycles per execution. When two instances of each atom type are available, only 18 cycles are required per execution. Note that a second instance of the QSub atom does not provide any benefit as the QSub calculation is actually hidden in the memory access latency and thus parallelizing the QSub calculations does not improve the performance. The fastest implementation is reached when for all atom types (except QSub) four instances are loaded (13 atoms altogether). This fastest implementation has rather large hardware requirement and is not necessarily beneficial in all situations. However, the concept of modular SIs allows dynamically choosing any particular molecule between cISA execution and the largest hardware molecule during run time and it supports gradually upgrading the SI implementation until the selected molecule is available.

In addition to the hardware molecules and the cISA molecule, so-called mixed molecules may also implement an SI (see Table 3.1). The term "mixed" denotes that some parts execute in hardware (using the available atoms) and some execute in software (using the cISA). A mixed molecule is a particular cISA implementation that additionally uses some atoms that already finished reconfiguration. Its purpose is to bridge the gap[4] between the cISA implementation (319 cycles; 0 atoms) and the smallest hardware molecule (22 cycles; 4 atoms). The cISA implementation of the mixed molecule executes a special assembler instruction that allows utilizing one atom at a time (i.e., any of the atoms that are currently available).

---

[4]In terms of molecule performance and atom requirements.

Table 3.1 Overview of different SATD molecule alternatives

Molecule type	Number of QSub atoms	Number of Repack atoms	Number of Transform atoms	Number of SAV atoms	Molecule latency [cycles]
cISA	–	–	–	–	319
Mixed	–	1	–	–	261
Mixed	–	1	1	–	173
Mixed	–	1	1	1	93
Hardware	1	1	1	1	22
Hardware	1	1	2	1	21
Hardware	1	2	2	1	20
Hardware	1	2	2	2	18
Hardware	1	4	4	4	16

This means that no molecule-level parallelism is exploited (at most one atom executes at a time), but it already benefits from the atom-level parallelism. Section 5.2 provides further details about implementing mixed molecules.

The Transform atom in Fig. 3.3 demonstrates an important feature of the proposed hierarchical SI composition. In addition to the two 32-bit inputs, it receives three control signals, namely IDCT (inverse DCT), DCT (discrete cosine transformation), and HT (Hadamard transformation). These control signals allow using this atom for the horizontal Hadamard transformation (left transform column in the SI graph in Fig. 3.3) as well as the vertical Hadamard transformation (right column) by applying different values to the HT control bit. In addition to the Hadamard transformation, the control bits allow using this atom to execute the SIs for DCT, inverse DCT, and inverse HT. Therefore, this atom can be shared between different SIs and different molecules of the same SI. This can cut down the actual area requirements, as potentially fewer atoms need to be loaded, depending on how many atoms can be shared. In addition, it reduces the reconfiguration overhead. For instance, if one Transform atom finishes reconfiguration then not only the SATD SI is expedited (a faster molecule is available), but potentially further SIs are also upgraded to faster molecules. Without atom sharing, each SI would need to reconfigure its own atoms to realize faster molecules.

As mentioned beforehand, a molecule implies resource constraints (type and quantity of available atoms) that affect the schedule of the SI graph. To illustrate this aspect further, Fig. 3.4 shows an excerpt of a schedule for the SATD SI (see Fig. 3.3) that utilizes two instances of each demanded atom type. Due to data dependencies, it is not always possible to utilize the available resources maximally, e.g., cycles 12 and 13 in Fig. 3.4. Therefore, intermediate results might need to be stored temporary, depending on the SI schedule. In addition, the communications between the atom instances need to be established to realize the molecule implementation. To be able to implement molecules, a computation and communication infrastructure was developed in the scope of the work that is presented in this monograph. It is a nonmodifiable structure that provides ACs and that can be extended by reconfiguring the ACs. This so-called atom infrastructure establishes the demanded

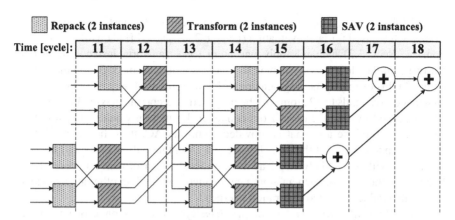

Fig. 3.4 Example schedule for a molecule of the SATD special instruction, using two instances of each atom

communications between the ACs and stores intermediate data from the atoms. Section 5.4 presents the details for the atom infrastructure.

In Sect. 3.1, this monograph identified three problems that state-of-the-art reconfigurable processors face, i.e., the sharing problem, fragmentation problem, and reconfiguration overhead problem. Now, the impact of the novel concept of modular SIs for these three problems is analyzed.

Reduced sharing problem. An atom can be shared between different molecules of the same SI as well as between different SIs. This is a significant improvement in comparison to monolithic SIs where sharing was disadvantageous (see Sect. 3.1). This improvement is practically usable, as, for example, demonstrated with the Transform atom that can be used to accelerate the SIs for (inverse) DCT and (inverse) Hadamard transformation. However, the concept of modular SIs does not completely solve the sharing problem. Sharing a partial data path within an atom (i.e., not the entire data path of an atom) faces the same challenges as sharing a data path within a monolithic SI. Conceptually, the size of an atom (i.e., the amount of logic that it contains) could be reduced to improve the potential for sharing relatively small data paths. However, this implies an increased overhead[5] to connect the atoms to compose molecules. Therefore, it corresponds to a trade-off between sharing potential and implied overhead. Even though this trade-off is independent of the underlying concept, it was considered, when designing the SIs, Molecules, and Atoms (see Sect. 3.3) for later evaluations.

Reduced fragmentation problem. Due to the partitioning of relatively large monolithic SIs into relatively small atoms, the fragmentation problem is reduced correspondingly. In the proposed concept, the maximal fragmentation is independent

---

[5]More atoms would need to interact (because there is less functionality per atom) and thus the hardware that connects the atoms would need to be extended to exchange more inter-atom data in parallel.

of the Molecule size, which is a noticeable improvement. For instance, the smallest SI that will be used for later evaluation (i.e., the 2 × 2 pixels' Hadamard transformation, see Sect. 3.3) demands only one atom (one transform instance; only one hardware molecule exists), whereas the largest molecule of SATD demands 13 atoms. In this example, the fragmentation of a monolithic SI may be as large as the area that corresponds to 12 atoms. In the proposed concept, the maximal fragmentation is limited by the size of an atom container. If an atom would demand only a small fraction of the area that is provided within an AC, then the fragmentation might be up to the size of the AC. However, an atom that only uses a small fraction of an AC will most likely not lead to a noticeably performance improvement. Potentially, the logic of such an atom may be incorporated into other atoms, which would also reduce the reconfiguration overhead.

Reduced reconfiguration overhead problem. Typically, as soon as one atom finished reconfiguration, a mixed molecule or hardware molecule is available to accelerate the SI execution. The reconfiguration time of one atom is approximately 1 ms (see Table 5.7) for the atoms that are implemented for the RISPP hardware prototype (see Sect. 5.5). The reconfiguration time for the fastest SATD implementation is approximately one order of magnitude larger (demanding 13 atoms). This significantly reduced reconfiguration overhead leads to significant performance improvements, as shown later. Furthermore, whenever an Atom can be shared between multiple SIs executing in the same hot spot, it only needs to be reconfigured once, which directly reduces the reconfiguration overhead. In addition, if an Atom can be shared between the SIs of different hot spots H_1 and H_2, it might be already available when H_2 demands it, because it was previously reconfigured for H_1. This eliminates the need to reconfigure this particular atom and thus it significantly reduces the reconfiguration overhead.

Altogether, the concept of modular SIs reduces the negative effects of the sharing problem, the fragmentation problem, and the reconfiguration overhead problem noticeably in comparison to monolithic SIs. In comparison to state-of-the-art ASIPs, these three aspects may still affect the performance but not as noticeably as state-of-the-art reconfigurable processors are affected. Vice versa, the adaptivity of reconfigurable processors provides the potential to outperform ASIPs, depending on the application requirements. If an application demands relatively few SIs and accelerators and if it demands always the same accelerators (e.g., because the application is dominated by a single hot spot, like the JPEG encoder or the ADPCM audio encoder, Rijndael encryption, etc.), then the adaptivity of reconfigurable processors might not provide significant benefit. However, if the application demands rather many SIs in multiple hot spots or if multiple tasks shall execute and these tasks altogether demand many SIs, then the adaptivity of state-of-the-art reconfigurable processors may outperform statically optimized ASIPs. In addition to addressing the problems of reconfigurable processors, the concept of modular SI also provides further advantages concerning the provided performance and adaptivity.

Efficient SI upgrades. In addition to solving the reconfiguration overhead problem (i.e., reducing the reconfiguration time until the SI execution is accelerated), modular

SIs also provide the possibility to upgrade the performance of the SIs further. By continuing reconfiguring atoms that are beneficial for a particular SI, faster molecules of that SI become available one after the other and thus the performance of the SI is gradually upgraded.

SI implementation adaptivity. Providing multiple implementations (i.e., molecules) per SI provides a very high level of adaptivity. It is no longer required to determine at compile time, which implementation shall be used to implement an SI. Instead, this decision can be dynamically adapted at run time, depending on the requirements of the application and depending on the hardware availability. The hardware availability might change at run time in a multitasking scenario. Depending on the amount of tasks, their priorities, deadlines, and hardware requirements (i.e., how much ACs they demand to be able to fulfill the performance requirements), the amount of AC that are available for a particular task might change. Even for a fixed number of ACs, the SI requirements of an application may change. For instance, it may depend on application input data, which SI is executed how often. If two SIs, SI_1 and SI_2, are executed several times in a hot spot, then – depending on input data – sometimes SI_1 might be executing more often and sometimes SI_2 (a concrete example for this is shown in Sect. 3.3). Therefore, depending on these SI execution frequencies it may be beneficial to adapt the SI implementation accordingly, i.e., spend more ACs for SI_1 or SI_2, respectively.

The concept of modular SIs diminishes the described problems of state-of-the-art reconfigurable processors and provides further potential and adaptivity. State-of-the-art monolithic SIs are actually a special case of modular SIs, i.e., they provide exactly one hardware molecule (with a predetermined performance/area trade-off) that is implemented using one dedicated rather large atom. However, to obtain the full benefits of modular SIs, multiple molecules per SI – composed of multiple shareable atoms – need to be provided. In addition, a hardware infrastructure is required that allows implementing modular SIs and furthermore a run-time system is needs to determine which molecule shall be used to implement an SI at a particular point in time. After summarizing the developed SIs, molecules, and SIs in the next section and providing a formal model of them in Sect. 3.4, this monograph will present an overview of the RISPP architecture in Sect. 4.1 (details in Chap. 5) and then explain and evaluate the RISPP run-time system in the remainder of Chap. 4.

3.3   Example Special Instructions for the ITU-T H.264 Video Encoder Application

In this section, an overview of the ITU-T H.264 video encoder [ITU05] is presented together with the designed SIs and atoms for acceleration. In addition, the complexity and the adaptive nature of this application are presented that makes it challenging for state-of-the-art application-specific processors (ASIPs) and reconfigurable processors. In subsequent chapters, this application is used to motivate and evaluate the components of the run-time system and of the entire RISPP approach. The H.264 encoder is significantly more complex than the applications from the MiBench [GRE+01] and the MediaBench [LPMS97] benchmark suites, which typically consist of one dedicated kernel[6] (e.g., a DCT kernel for the JPEG encoder, a filtering kernel for edge detection, or a Rijndael kernel for cryptography). If an application only contains a single kernel, i.e., only one or few SIs are required and they all execute in the same inner loop, then an ASIP might provide sufficient performance and efficiency for this particular application. In such a single-kernel application, reconfigurable processors only demand a single reconfiguration when the application starts (or even before it starts) to load the implementations of the SIs. After that reconfiguration, the reconfigurable processor operates similar to an ASIP. The reason for this single reconfiguration is the time between the SI executions. This is due to the fact, that all SIs are part of the same kernel and the time between their execution is too small to perform a reconfiguration. Still, many reconfigurable projects (e.g., Warp [LSV06], Molen [VWG+04], Proteus [Dal03], OneChip [CC01]) use these applications for evaluation, which diminishes the effects of the reconfiguration overhead, because no reconfigurations during run time are demanded to accelerate single-kernel applications. Even the initial overhead of an online synthesis (like for the Warp processor) may amortize if the application executes long enough.

The application scenario becomes more challenging if multiple SIs are demanded in different kernels or different computational blocks and – due to area constraints or adaptivity aspects – run-time reconfiguration is required to support them. These multiple SIs may be required by one or by multiple applications (multitasking scenario). The H.264 video encoder is such an example that requires multiple SIs. In addition, it has a high computational complexity (~10× relative to MPEG-4 simple profile encoding [OBL+04]) and – when for instance used for video conferencing – challenging timing constraints (33 ms for 1 frame audio/video en/decoding when targeting 30 frames per second).

Figure 3.5 provides an overview of the H.264 video encoder application flow. It consists of three computational blocks that are executed subsequently for each video frame. All three internally iterate over all MacroBlocks (MBs, 16 × 16 pixel blocks) of the frame. The first computational block performs the motion estimation (ME) to exploit temporal redundancies in two consecutive video frames, i.e., for each MB of the current frame it tries to find a similar MB in the previous frame. If such a matching MB is found, encoding the MB in the current frame can be done by storing a motion vector and the differences of this MB to the MB in the previous frame to which the motion vector points. After the motion estimation, the rate distortion optimization (RDO) decides which MB prediction type and which

---

[6]Please note that this thesis distinguishes between a computational block (e.g., the motion estimation of an H.264 video encoder) and a kernel (a part of a computational block, e.g., sum of absolute differences (SAD) and sum of absolute transformed differences (SATD) for the motion estimation); typically, a computational block corresponds to the outer loop of a computation and the kernel corresponds to the inner loops.

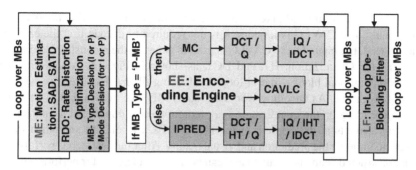

Fig. 3.5 H.264 application flow, highlighting the three main computational blocks (ME, EE, LF) and their embedded kernels

corresponding mode shall be used for later encoding (considering the expected quality (i.e., distortion) and bit rate of the resulting video). For instance, for each MB one out of two fundamentally different MB prediction types (intraframe and interframe) is chosen. The interframe prediction uses a motion vector to predict the content of an MB in the current frame (so-called P-MB) from an MB in the previous frame. The intraframe prediction uses some of the surrounding pixels of the MB in the current frame (so-called I-MB) to predict its content. For the I-MB, the mode also decides which surrounding pixels shall be considered in which way.

The encoding engine (EE) is the second computational block. It performs the actual prediction (decided by ME and RDO) and encoding for all MBs, depending on their MB types, i.e., P-MB or I-MB (see "then" and "else" path in Fig. 3.5). The encoding comprises computing a discrete cosine transformation (DCT), a Hadamard transformation (HT, depending on the MB type), and quantizing the result. Afterwards, the mode decisions and encoded data are processed using a context-adaptive variable-length encoding (CAVLC, a lossless entropy coding) in order to generate an encoded bitstream. At the encoder side, the encoded frame additionally needs to be decoded in order to generate a reconstructed frame (so-called reference frame) for the ME of subsequent input video frames (i.e., for P-MBs). Moreover, for I-MB prediction, the prediction data is generated from the pixels of neighboring reconstructed MBs. Creating this reconstructed data at the encoder side is required to be able to successfully decode the data at the decoder side. The reason is that the decoder has no knowledge of the input video frame to the encoder and only the pixel difference between current MB and the prediction data (e.g., reference MBs of the previous frame in case of P-MB) is available in the encoded bitstream (which is transmitted to the decoding side). To decode a P-MB, the pixel difference is added to the reference MB of the previous frame. If the decoder uses different data for the reference MB than the encoder did, then the decoded result will differ. As the decoded MBs will be used as reference to decode the next video frame, this difference increases from frame to frame. Therefore, to assure that the encoder and decoder use the same reference data, the encoder has to use the reconstructed video data (i.e., decoded) as reference instead of the original input video frame (i.e., not encoded). For this reason, a partial decoder is embedded in the encoder that generates the reconstructed pixel data similar to a decoder generating the decoded pixel data. Since CAVLC is lossless, it is not required to decode it at the encoder side. Instead, the input data to CAVLC is used to generate the reconstructed data by inversing the Quantization, HT, and DCT, as shown in Fig. 3.5. Afterwards, the third computational block – the in-loop de-blocking filter (LF) – iterates over the MBs of the frame. As encoding works on (sub-) MB level (down to 4 × 4 pixels) and – depending on the quantization – a prediction error might lead to noticeable blocking artifacts at the borders of the (sub-) MBs. LF aims to smoothen these borders to improve the prediction and visual quality. After this last step is performed, the reconstructed frame is obtained and can be used as reference for performing ME on the next frame.

One remarkable property of this video encoding process is the fact that the computational requirements may differ significantly during run time. This is because various different encoding modes and types exist for the MBs and it depends on certain properties – that are not known at compile time – which MB type and mode will be chosen for a frame. In particular, it depends on input video characteristics, the targeted bit rate, and quality. Figure 3.6 shows an analysis of a video sequence after its encoding was performed. The relative number of I-MBs (in percent) is shown for each frame. The relative number of P-MBs varies directly opposed to the curve. In the first 200 frames, only around 10% of the MBs (in this video resolution 40 MBs) are encoded as I-MBs. This is because the video sequence was showing a relatively smooth scene where the motion estimation was able to detect good reference MBs, i.e., the P-MBs where more beneficial. Between 250 and 350 frames, the scenes contain rather hectic motion that makes it hard for the motion estimation to find a corresponding MB in the previous frame. Therefore, the distribution of I-/P-MBs change, as now the intraframe prediction became more beneficial and

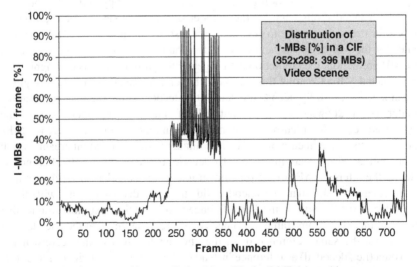

Fig. 3.6 Relative frequency of intraprediction MacroBlocks (I-MBs) in a video sequence

thus the relative number of I-MBs increases. This distribution of I-/P-MBs is not generally predictable during compile time if the input video sequence is not known in advance. As shown in Fig. 3.5, the computational requirements for I-MBs and P-MBs are very different. Therefore, for changing I-/P-MB distribution, the corresponding kernels (motion compensation, intraprediction, etc.) will be demanded in different frequencies. Thus, it is very hard to predetermine which share of the available hardware shall be used to accelerate the kernels for I-MB encoding and which shall be used for the P-MBs. Both, ASIPs and state-of-the-art reconfigurable processors have to make a compromise here (i.e., provide a statically determined share of the fabric to I-MB and P-MB, respectively). ASIPs make this decision at design time and reconfigurable processors at compile time. During a run-time phase where – for instance – most encoding is performed with P-MBs, the accelerators for the I-MB are rather idle and do not contribute to the performance or the efficiency of the encoding in a positive way. This is one of the examples (as described in Sect. 3.2) where the extended SI implementation adaptivity of the RISPP approach can be used to adapt the available reconfigurable fabric to the demanded kernels.

To accelerate the kernels in Fig. 3.5 multiple modular SI were implemented, for instance the SATD example in Fig. 3.3 accelerates the motion estimation. Figure 3.7 shows the SI for the motion compensation (used to encode P-MBs) together with the point filter atom and the Clip3 atom [SBH09a]. The point filter implements a data path that performs eight 8-bit multiplications in parallel and accumulates all results. The Clip3 atom modifies the input values and implements a control flow that restricts the values to fixed upper and lower limits. This shows that atoms may implement data flow, control flow, as well as combined operations.

Figure 3.8 shows a loop filter SI [SBH09a]. The operation contains a condition calculation and a filtering operation that is based on the determined conditions. The SI is composed of four independent Condition–Filter pairs as shown in Fig. 3.8a. Figure 3.8b shows a schedule for the molecule that comprises two instances of the condition atom and two instances of the filter atom. As shown in part a, a condition–filter pair obtains two 32-bit pixel values that shall be filtered and two 32-bit threshold values to determine the conditions (additionally, control signals are applied that for instance allow to reuse the condition atom for Luma (brightness component) and Chroma (color component) of a frame). Altogether, one condition–filter pair obtains 128-bit input data (in addition to the static control signals) that may be loaded from data memory in a single access if a memory port with 128 bit is available. For the RISPP approach, two 128-bit data memory ports are used that access fast on-chip memory (like Tensilica ASIPs support it [Tenb]) as mentioned in Sect. 4.1 and explained more detailed in Sect. 5.3. Correspondingly, the input data for two condition–filter pairs can be loaded in parallel, as shown in Fig. 3.8b. When additionally considering that fast on-chip memory is used that guarantees a memory access latency of two cycles, the load and store operations of four condition–filter pairs can be efficiently interleaved such that the full available data memory bandwidth is exploited. This explains why it is beneficial to combine these four independent condition–filter pairs into the same SI instead of offering a single pair as SI. It demonstrates that the SI design – to some degree – depends on

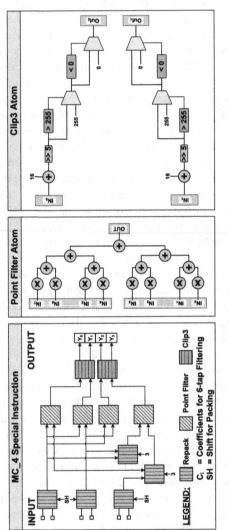

Fig. 3.7 Example of a motion compensation special instruction with three different atoms and the internal data path of the point filter atom

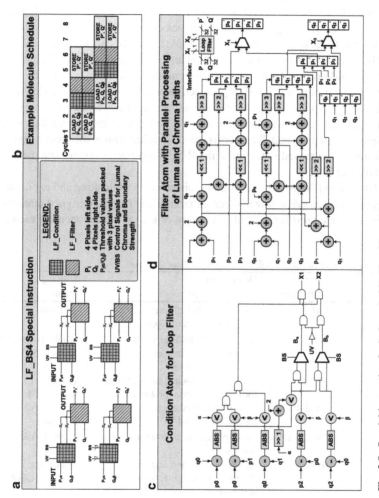

Fig. 3.8 Special instruction for in-loop de-blocking filter with example schedule and constituting atoms for filtering conditions and filtering operation

the underlying architecture. If only one 128-bit port would be available, then only two pairs within one SI would already exploit the maximally available bandwidth.

Figure 3.8c, d shows the internal composition of the condition- and filter atom of the loop filter SI. It is especially noticeable that very fine-grained (i.e., subbyte to bit level) computations are performed. For instance, the operations in the condition atom go down to elementary *AND* gates. In addition to various byte and subword operations, the filter atom also comprises multiple shift operations with a constant shift amount. Therefore, the operation is not realized as an actual shifting but as a simple bit-level rewiring (e.g., input_bit$_i$ becomes output_bit$_{i+1}$).

In addition to the above-described SIs, multiple further SIs are used to accelerate the kernels of the H.264 video encoder. Table 3.2 summarizes all of these SIs [SBH09a]. In addition to the name and the computational block where the SI is used, a short description for the SI is provided as well as a list of demanded atom types. Each SI (except HT_2 × 2 that only requires one instance of one atom) comprises multiple mixed and hardware molecules. It becomes noticeable that some atoms (especially Repack and Transform) are used to implement different SI types. This indicates the increased potential of the RISPP approach to share data paths as described in Sect. 3.2.

Note. The SIs and Atoms that are used in this work for motivating and explaining the concepts as well as evaluating and benchmarking them are all designed manually [SBH09a]. They are used for a proof of concept of the core components presented

Table 3.2 Overview of implemented SIs and their required atoms

Functional component	Special instruction	Description of special instructions	Utilized atoms
Motion estimation	SAD	Sum of absolute differences of a 16 × 16 MacroBlock	SADrow
	SATD	Sum of absolute (Hadamard-) transformed differences of a 4 × 4 subblock	QSub, Transform, Repack, SAV
Motion compensation	MC_Hz_4	Motion compensated interpolation for horizontal case for 4 pixels	PointFilter, Repack, Clip3
Intraprediction	IPred_HDC	16 × 16 Intraprediction for horizontal and DC	CollapseAdd, Repack
	IPred_VDC	16 × 16 intra prediction for vertical and DC	CollapseAdd, Repack
(Inverse) Transform	(I)DCT	Residue calculation and (inverse) discrete cosine transf. for 4 × 4 subblock	Transform, Repack, (QSub)
	(I)HT_2 × 2	2 × 2 (Inverse) Hadamard transform of Chroma DC coefficients	Transform
	(I)HT_4× 4	4 × 4 (Inverse) Hadamard transform of intra-DC coefficients	Transform, Repack
In-loop de-blocking filter	LF_BS4	4-Pixel edge filtering for in-loop de-blocking filter with boundary strength 4	Cond, LF_4

in this monograph, i.e., the RISPP architecture and the RISPP run-time system. An automatic SI detection is not targeted in this work. However, automatic SI detection and synthesis is an established research field in the context of ASIPs (see Sect. 2.1). To automatically partition an SI into reusable atoms, techniques like data path sharing may be used. The research project KAHRISMA [ITI, KBS+10] that – to some degree – builds upon the concepts presented in this monograph investigates the automatic detection and partitioning of modular SIs further.

3.4 Formal Representation and Combination of Modular Special Instructions

In this section, this monograph explains the hierarchical SI composition on a formal basis to present the molecule/atom interdependencies and to simplify and clarify expressions in the pseudo code in Chap. 4 later on. Let us define a data structure $(\mathbb{N}^n, \cup, \cap)$, where $\mathbb{N}^n$ contains all molecules and n is the number of all defined atoms types. A special instruction is represented as a set of molecules. For expediency, consider $\vec{m}, \vec{o}, \vec{p} \in \mathbb{N}^n$ as molecules with $\vec{m} = (m_0, \ldots, m_{n-1})$ where m_i describes the desired number of instances of atom type A_i to implement the molecule (similarly for $\vec{o}$ and $\vec{p}$). The operators $\cup$ and $\cap$ are used to combine two molecules and create a so-called meta-molecule (i.e., a combination of atoms that does not necessarily correspond to an implementation of an SI). The operator $\cup$ (see Eq. 3.1) describes a meta-molecule that contains the atoms that are required to implement both molecules of its input. The operator $\cap$ (see Eq. 3.2) describes a meta-molecule that contains the atoms that are demanded by both molecules of its input (i.e., that can be shared by both molecules). Since the operator $\cup$ is commutative and associative with the neutral element $(0, \ldots, 0)$, therefore $(\mathbb{N}^n, \cup)$ is an Abelian semigroup. The same is true for $(\mathbb{N}^n, \cap)$ with the neutral element $(maxInt, \ldots, maxInt)$. The determinant of a molecule is defined as the number of atoms it comprises (see Eq. 3.3). Figure 3.9 provides an example for these three operators. To simplify the visualization, let us restrict n (i.e., the number of different atom types) to two. The figure shows two example molecules $\vec{o}$ (left upper circle) and $\vec{p}$ (right lower triangle) with their corresponding atom requirements that is indicated by the correspondingly colored rectangles. For instance, the rectangle for $\vec{o}$ reaches from $(0, 0)$ to $(1, 4)$ and covers all potential atom combinations that are available if the atoms for $\vec{o}$ are available. This means that any other molecule within that rectangle is physically available (i.e., all of its atoms are reconfigured) if $\vec{o}$ is available.

$$\cup : \mathbb{N}^n \times \mathbb{N}^n \to \mathbb{N}^n; \vec{o} \cup \vec{p} := \vec{m}; \; m_i := \max\{o_i, p_i\} \tag{3.1}$$

$$\cap : \mathbb{N}^n \times \mathbb{N}^n \to \mathbb{N}^n; \vec{o} \cap \vec{p} := \vec{m}; \; m_i := \min\{o_i, p_i\} \tag{3.2}$$

$$| \; | : \mathbb{N}^n \to \mathbb{N}; \; |\vec{m}| := \sum_{i=0}^{n-1} m_i \tag{3.3}$$

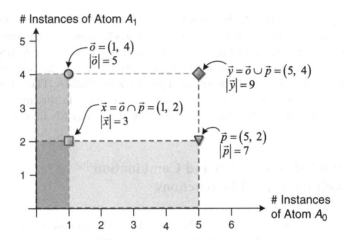

Fig. 3.9 Example for union, intersection, and determinant operation on molecules

$$\triangleright: \mathbb{N}^n \times \mathbb{N}^n \to \mathbb{N}^n; \vec{o} \triangleright \vec{p} := \vec{m}; \ m_i := \begin{cases} p_i - o_i, if \ p_i - o_i \geq 0 \\ 0, \text{else} \end{cases} \tag{3.4}$$

$$/: \mathbb{N}^n \times \mathbb{N}^n \to \mathbb{N}^n; \ \vec{p}/\vec{o} := \vec{o} \triangleright \vec{p} \tag{3.5}$$

In addition to these elementary operations, further functions that allow representing the concept of molecule upgrading (as introduced in Sect. 3.2) are defined. For instance, some of the later-described algorithms need to know which atoms are additionally required to realize a molecule $\vec{p}$ when the atoms of molecule $\vec{o}$ are already available. The upgrade operator $\triangleright$ is used to represent these additionally required atoms (see Eq. 3.4). Figure 3.10 provides two examples for this operator. To upgrade from $\vec{o}_1$ to $\vec{p}$ one instance of atom A_0 and two instances of atom A_1 are demanded. This is denoted by the grey box going from $\vec{o}_1$ to $\vec{p}$. The resulting meta-molecule that represents $\vec{o}_1 \triangleright \vec{p}$ corresponds to the position of $\vec{p}$ in the grey box (i.e., the upper right corner) if the position of $\vec{o}_1$ in the box (i.e., the lower left corner) is transposed to the origin $(0, 0)$ as indicated in the figure. The second example – upgrading from $\vec{o}_2$ to $\vec{p}$ – shows that the upgrade destination does not necessarily demand more atom instances for all atom types. In this case, the box may degenerate to a line or even a point if no additional atoms are required. A variation of the upgrade operator is the relative complement calculation (i.e., difference). It can be used to determine which atoms are left "without" the atoms of a molecule $\vec{o}$ if the atoms of $\vec{p}$ are available. It becomes apparent, that this operation directly corresponds to an upgrade from $\vec{o}$ to $\vec{p}$ (see Eq. 3.5). For instance, if the atoms for molecule $\vec{p}$ in Fig. 3.10 are available and the atoms for $\vec{o}_1$ shall be removed (i.e., $\vec{p}$ without $\vec{o}_1$), then exactly the atoms $\vec{o}_1 \triangleright \vec{p}$ (indicated by the grey box) remain. In addition to combining molecules – using the introduced operators – compare them is required. Therefore, let us define the relation $\vec{o} \leq \vec{p}$ as shown in Eq. 3.6. As the relation is reflexive, antisymmetric, and transitive, $(\mathbb{N}^n, \leq)$ is a partially

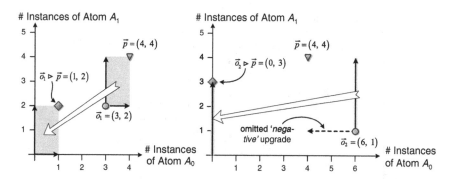

Fig. 3.10 Example for upgrade operation on molecules

ordered set. For a set of molecules $M \subset \mathbb{N}^n$, a supremum and an infimum is defined as shown in Eqs. 3.7 and 3.8, respectively. The supremum from M is a meta-molecule with the meaning of declaring all atoms that are needed to implement any of the molecules in M, i.e., $\forall \vec{m} \in M : \vec{m} \leq \sup(M)$. The infimum is correspondingly defined to contain those atoms that are collectively needed for all molecules of M. $\varnothing \neq M \subset \mathbb{N}^n$ has a well-defined supremum and infimum, $(\mathbb{N}^n, \leq)$ is a complete lattice. Figure 3.11 provides an example for the relations of six different molecules $M = \{\vec{o}_1, \dots, \vec{o}_6\}$ including their supremum and infimum. The "relation" lines that connect the supremum and infimum are omitted for clarity. It is noticeable that not all molecules are in relation to each other, for instance, $\vec{o}_6$ is only in relation to the supremum and infimum. In comparison to $\vec{o}_4$ it is not smaller or equal than $\vec{o}_6$ (i.e., $\vec{o}_6 \not\leq \vec{o}_4$), because it has more instances of atom A_1. Vice versa, $\vec{o}_4$ has more instances of A_0 than $\vec{o}_6$, thus $\vec{o}_4 \not\leq \vec{o}_6$. Furthermore, it is noticeable, that neither the supremum nor the infimum are a member of M; therefore, it is clear that this relation does not always have a maximum (i.e., $\sup(M) \in M$) or a minimum (i.e., $\inf(M) \in M$).

$$\vec{o} \leq \vec{p} := \begin{cases} \text{true, if } \forall i \in [1, n] : o_i \leq p_i \\ \text{false, else} \end{cases} \tag{3.6}$$

$$\sup : \mathcal{P}(\mathbb{N}^n) \to \mathbb{N}^n; \ \sup(M) := \bigcup_{\vec{m} \in M} \vec{m} \tag{3.7}$$

$$\inf : \mathcal{P}(\mathbb{N}^n) \to \mathbb{N}^n; \ \inf(M) := \bigcap_{\vec{m} \in M} \vec{m} \tag{3.8}$$

To obtain an explicit access to individual atoms and their quantity in a molecule, let us write a molecule as a multiset, i.e., a set that may contain an element multiple times. Equation 3.9 shows how the vector notation of a molecule $\vec{m}$ can be transformed into a multiset notation $\tilde{m}$ where the elements correspond to atoms A_i with quantity m_i. To be able to apply the previously defined operators on these elements (i.e., the atoms), let us define a convenience transformation T (see Eq. 3.10) that allows us to convert an atom into a vector representation (i.e., a molecule that

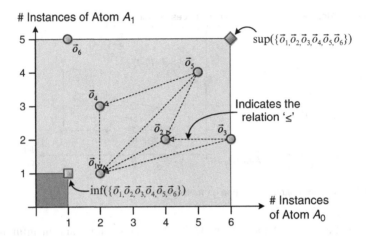

Fig. 3.11 Example for relation, supremum, and infimum of molecules

contains exactly one instance of this atom). The operator for addition (see Eq. 3.11) allows us to combine multiple transformed atom instances to be able to represent a molecule as a linear combination of its composing atoms. Equation 3.12 shows such a linear combination for molecule $\tilde{m}$, where all atom instances are transformed into a vector representation, summed up in the corresponding quantity (inner loop), and then added to the corresponding sums of all atom types (outer loop), eventually leading to the vector representation $\vec{m}$.

$$\vec{m} = (m_0, \ldots, m_{n-1}) \rightarrow \tilde{m} := \left\{ \underbrace{A_0, \ldots, A_0}_{m_0}, \ldots, \underbrace{A_{n-1}, \ldots, A_{n-1}}_{m_{n-1}} \right\} \tag{3.9}$$

$$T : \{A_i\} \rightarrow \mathbb{N}^n ; \ T(A_i) := (0, \ldots, 0, \underbrace{1,}_{i^{\text{th}} \text{entry}} 0, \ldots, 0) \tag{3.10}$$

$$+ : \mathbb{N}^n \times \mathbb{N}^n \rightarrow \mathbb{N}^n ; \ \vec{o} + \vec{p} := \vec{m}; \ m_i := o_i + p_i \tag{3.11}$$

$$\tilde{m} = \left\{ \underbrace{A_0, \ldots, A_0}_{m_0}, \ldots, \underbrace{A_{n-1}, \ldots, A_{n-1}}_{m_{n-1}} \right\} \rightarrow \vec{m} = \sum_{i=0}^{n-1} \sum_{j=0}^{m_i} T(A_i) \tag{3.12}$$

In addition to the formal model for atoms and molecules, the algorithms of the run-time system (see Chap. 4) demand extra information that is not specific to the hierarchical SI composition. For instance, it is important to know how fast a certain molecule is executed (i.e., its latency) irrespective of which atoms are demanded to implement that molecule or whether these atoms may be shared with other molecules. To provide this information dedicated functions are used that can be "called"

Table 3.3 Overview of high-level molecule and special instruction properties

Operator for molecule $\vec{m}$ or special instruction s	Description
int $l = \vec{m}.getLatency()$	A molecule has certain execution latency (in cycles). For some of the later-described algorithms this is an important information that can be obtained using this function
$s = \vec{m}.getSI()$	Some information is specific to an SI (i.e., they are identical to all molecules of that SI) and thus are maintained as part of the SI. This function allows to get access to the SI of a molecule and thus to the corresponding information. Note, for a so-called meta-molecule (i.e., a vector of atoms that does not directly implement any particular SI) the value NULL is returned
$\vec{m} = s.getCISAMolecule()$	Returns that molecule of the SI that uses the cISA to execute the SI, i.e., $\vec{m} = (0,\ldots,0)$, providing access to the cISA latency
$\vec{m} = s.getFastestAvail-\ ableMolecule(\vec{a})$	This function is a convenience function for the following operation: $\vec{m} : \vec{m}.getLatency() =$ $\min\left\{\vec{o}.getLatency() \mid \vec{o}.getSI() = s \wedge \vec{o} \le \vec{a}\right\}$ It returns the molecule $\vec{m}$ that provides the fastest latency for SI s and that can be implemented with the atoms described by the meta-molecule $\vec{a}$
int $f = s.getExpected-\ Executions()$	To execute a certain computational block, its SIs are executed in a certain frequency. This frequency might depend upon input data as described in Sect. 3.3. This function provides an estimation on the expected execution frequency of a particular SI. This frequency is estimated by an online-monitoring approach (see Sect. 4.3) and used to determine which molecule shall be reconfigured to execute an SI (see Sect. 4.4).

for a particular molecule or SI and that provide these specific parameters. The syntax is oriented at the widely spread object-oriented programming style, i.e., a molecule can be seen as an object and the function is called for that particular object, potentially providing parameters to guide the function. For instance, " $\vec{m}.getLatency()$ " would provide the execution latency for molecule $\vec{m}$. Table 3.3 provides an overview of all these high-level properties (i.e., not directly related to atom/molecule composition) with their syntax and an explanation.

3.5 Summary of Modular Special Instructions

This chapter has identified and discussed three major problems of state-of-the-art reconfigurable processors that offer their special instructions (SIs) as monolithic blocks, i.e., the sharing problem, fragmentation problem, and reconfiguration overhead problem. Afterwards, this monograph presented a novel hierarchical SI composition that allows providing novel modular SIs that are based on atoms and molecules. These modular SIs target the three identified problems and additionally provide efficient SI upgrading and extended SI implementation adaptivity. Therefore, the proposed modular SIs provide a high potential for adaptivity and efficiency. This potential is exploited by the novel RISPP run-time system (presented in the next chapter) to provide a high performance. An H.264 video encoder example application with multiple SIs and their composing atoms and molecules was presented in this chapter. It demonstrates the feasibility of the novel concept of modular SIs applied to real-world applications and will be used for later evaluation of the proposed concept and for comparisons of the RISPP approach with state-of-the-art processors. To put the proposed concept on a solid base, a formal model was developed that allows describing SIs, Molecules, and Atoms. This model allows combining molecules (using the defined operators) to describe the tasks of the run-time system and pseudo codes of the proposed algorithms in Chap. 4 in a more clear and precise way.

Chapter 4
The RISPP Run-Time System

This chapter presents the novel run-time system of the RISPP architecture. It exploits the novel concept of modular special instructions (SIs) [BSKH07], as described in Chap. 3. The first section will present a short overview of the RISPP architecture. The focus of that section is placed on describing those parts of the architecture that are required to understand the run-time system. The entire RISPP architecture including the novel computation and communication infrastructure [BSH08a] is described in Chap. 5. The second section analyses the requirements of the run-time system, provides a first overview of its tasks, and describes how these tasks interact, using a state-transition diagram [BSH08b].

Modular SIs provide different implementation alternatives for SIs (i.e., molecules) that are composed out of atoms. The run-time system needs to determine which SIs are demanded for a computational block to provide SI implementations for its acceleration. Section 4.3 describes, how online monitoring and a lightweight error back-propagation scheme are used to predict which SIs are demanded and how often they are expected to be executed [BSTH07]. Section 4.4 describes how this information is used to determine a molecule for each SI that is predicted to be executed in the upcoming computational block, such that the atoms that are required to implement all molecules fit to the available reconfigurable fabric [BSH08c]. Afterwards, Sect. 4.5 presents how the atom reconfiguration sequence is determined, i.e., which atom shall be reconfigured first, etc. [BSKH08]. Whenever a new reconfiguration shall be started, an existing atom might need to be replaced and in Sect. 4.6 the developed replacement policy is presented [BSH09b].

4.1 RISPP Architecture Overview

After introducing the novel concept of modular SIs in Chap. 3, this section provides an overview of the RISPP hardware architecture. Modular SIs provide the potential for a high adaptivity and for reducing the effects of the reconfiguration overhead, as it was described in Chap. 3 and as is evaluated in the following sections and in Chap. 6. However, to exploit the potential of modular SIs an enabling hardware architecture

L. Bauer and J. Henkel, *Run-time Adaptation for Reconfigurable Embedded Processors*,   55
DOI 10.1007/978-1-4419-7412-9_4, © Springer Science+Business Media, LLC 2011

and a supporting run-time system is demanded that – among others – decides, which molecule shall be provided at which point in time and in which sequence the demanded atoms shall be reconfigured. Before specifying the actual tasks of the run-time system in Sect. 4.2 and presenting the details in the following sections, this section will provide an overview of the underlying hardware architecture. In this section, the focus is placed on those parts of the RISPP architecture that are required to explain the run-time system afterwards. The full implementation details and prototyping results are given in Chap. 5.

Figure 4.1 provides an overview of a particular instance of the RISPP architecture. The RISPP concept can be embedded into different pipeline-based processors and is not limited to a specific one. In this work presented in this monograph, simulations, implementations, and experiments were performed on two different pipeline processors, i.e., a DLX processor from ASIPMeister [ASI] and the Leon2 processor (implementing the Sparc-V8 instruction set architecture) from Gaisler [Aer]. Figure 4.1 shows a simplified five-stage pipeline structure, comprising instruction fetch, instruction decode, instruction execute, memory access, and register write back stages. The required processor extensions to support modular SIs conglomerate in the execute stage and is explained in the following.

The *atom infrastructure* is connected to the execution stage as a functional unit (like the ALU). It is a fixed (i.e., not reconfigurable) part of the RISPP architecture. However, it contains partially reconfigurable regions – so-called atom containers (ACs, see rectangles inside the atom infrastructure in Fig. 4.1) – that can be dynamically

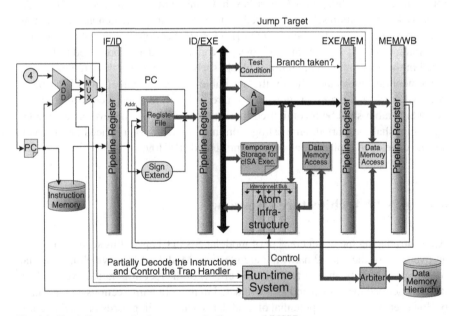

Fig. 4.1 Extending a standard processor pipeline toward RISPP

reconfigured (reconfiguration time of one AC is approximately 1 ms, see Table 5.7) to contain any particular atom without affecting the atom infrastructure or other ACs during reconfiguration. Therefore, the atom infrastructure itself is nonmodifiable, but it is extensible by loading atoms into the ACs. The ACs are coupled by an interconnect structure (multiple segmented busses, details are explained in Sect. 5.4), which allows that multiple atoms interact to realize a molecule, i.e., to implement the functionality of an SI. When the demanded atoms to implement an SI are not available, then the functionality of the SI is implemented using the core Instruction Set Architecture (cISA, as motivated in Chap. 3). This cISA implementation is triggered by a trap and the corresponding trap handler has to identify which SI shall execute and which input data shall be provided via the register file. To assist this initial data acquisition, the *temporary storage for cISA execution* (see Fig. 4.1) automatically stores this information and provides it to the trap handler (details are explained in Sect. 5.2).

In general, the input data for an SI is provided by the register file or data memory and the computational results of an SI are written to data memory or transferred to the register file via the write-back stage of the pipeline. The register file is extended to provide four read ports and two write ports to provide sufficient input/output data for the SI (details are explained in Sect. 5.1). The register file input can also be used to provide memory addresses (e.g., base address and strides of vectors; see Sect. 5.4.2) to be able to process larger amount of data in an SI (as, e.g., demanded for the SI examples in Figs. 3.3, 3.7, and 3.8). A *data memory access* unit realizes access to the data memory hierarchy (shared memory), providing two independent data memory access ports with 128 bit access width each. The *arbiter* decides whether the memory stage of the pipeline or the data memory access unit for the SIs is allowed to access the data memory whenever a conflicting access occurs, i.e., when both demand access at the same time.

When the memory stage accesses an address during the execution of an SI that also accesses the same address, then a potential memory inconsistency problem may occur. For instance, this problem was investigated in [CC01] and even though they provide a significant amount of hardware support to resolve all nine identified memory inconsistency problems automatically, only a minor performance improvement was reported. Therefore, RISPP stalls the pipeline during the execution of an SI. This minimizes the potential memory inconsistency problems and simplifies the arbiter implementation (details are explained in Sect. 5.3).

The *run-time system* (see Fig. 4.1) is controlling the atom infrastructure in two different ways. One the one hand it is determining the reconfiguration decisions and on the other hand it is controlling the SI executions. Determining the reconfiguration decisions directly affects the content of the atom containers in the atom infrastructure and thus it determines which molecules will be available for the different SIs. Controlling the SI executions depends on the available molecule, for instance, a trap may be triggered if the demanded atoms to implement an SI are not available or the interconnect structure in the atom infrastructure needs to be controlled to connect the available atoms in such a way that they realize the molecule's functionality.

4.1.1   Summary of the RISPP Architecture Overview

This section showed the fundamentals of the RISPP architecture (that allow realizing modular SIs) and presented an architecture overview (details are given in Chap. 5) that allows explaining the run-time system in the following sections. Without this run-time system, the major advantages of the novel modular SIs (i.e., adaptivity by changing the hardware overhead/performance trade-off at run time and upgrading the SI implementation to diminish the reconfiguration overhead) could not be utilized in an efficient and adaptive manner.

4.2   Requirement Analysis and Overview

Chapter 3 introduced the novel concepts of modular special instructions (SIs) and discussed their potential advantages in comparison to state-of-the-art approaches. Section 4.1 provided an overview of the RISPP architecture that allows realizing modular SIs without going into implementation details (provided in Chap. 5). This section will analyze which decisions are determined at design- and compile time, and which decisions need to be done during application run time. The basic idea is that only run-time decisions allow run-time adaptivity with its potential benefits. However, they also introduce a certain overhead that is demanded to determine the decisions. Therefore, a careful investigation is required to determine which decisions may be made at design and compile time and which decisions need to be made during run time. Afterwards, an overview of the run-time system is given, showing how it is connected to the RISPP architecture. Then, the tasks of the run-time system are described in more detail. The actual formal problem descriptions, implementation alternatives, and pseudo codes for the individual tasks of the run-time system are given in the following sections.

The concept of modular SIs provides an improved potential for adaptivity according to different SI implementation alternatives. However, neither the core pipeline of the architecture, nor the basic SI interface is affected by this concept, as the concept mainly targets the SI-internal composition. Therefore, similar to state-of-the-art reconfigurable processors, relatively large parts of the core pipeline may be fixed at design time (i.e., when the architecture is synthesized, placed and routed, and – in case of an ASIC design – taped out). For instance, the core pipeline including the number of pipeline stages, cISA, arithmetic and logic unit, and register file can be fixed at design time. The RISPP approach is not limited to any particular core pipeline (e.g., simulations and prototyping was performed for a DLX-based [HP96] and a SPARC-based [SPA] core pipeline); however, at design time one particular core pipeline needs to be determined and fixed. Similarly, some parts of the SIs need to be determined at design time. To be able to provide application-specific SIs after designing the CPU, not all parts of the SIs need to be fixed, for instance, it should be possible to add new SIs with new functionalities after design time. However, the instruction format, the input and output parameters of the SI,[1] and potential opcodes for the SIs can be fixed.[2] Similarly, the data memory access (number of ports and bit widths per port) for SIs can be fixed (see also Fig. 4.1 in the previous section). To some degree, these decisions may limit the SI, for instance, some SIs may benefit from a larger data memory access, whereas others would get along with a smaller data memory access than the chosen one. However, these decisions correspond to a typical design-space exploration to determine which feature set promises a certain performance at reasonable cost (e.g., area wise). The RISPP approach does not affect this design-space exploration significantly, i.e., modular SIs are functional with a relatively small data memory bandwidth as well as with a relatively large data memory bandwidth [BSH09a] (though providing more performance at a higher memory bandwidth). However, due to the concept of reconfigurable modular SIs, further parameters need to be decided during design time [BSH09a]. For instance, the size of the reconfigurable fabric (i.e., how many atoms may be available at the same time) is one of these parameters. In general, all parameters according the implementation of the atom infrastructure (see Sect. 4.1 and Sect. 5.4) need to be determined. Furthermore, the implementation of the run-time system needs to be determined to some degree (some parts may be left parametrizable, as shown later).

Figure 4.2 provides an overview of the steps to be done at design-, compile-, and run time. After the architectural parameters are determined at design time, the main compile-time decisions are specific to the SIs. At first, application-specific SIs need to be determined and corresponding atoms need to be designed. Defining the SIs comprises assigning an opcode (out of the design-time reserved opcodes), providing a cISA implementation and creating an SI graph (nodes are atoms and edges

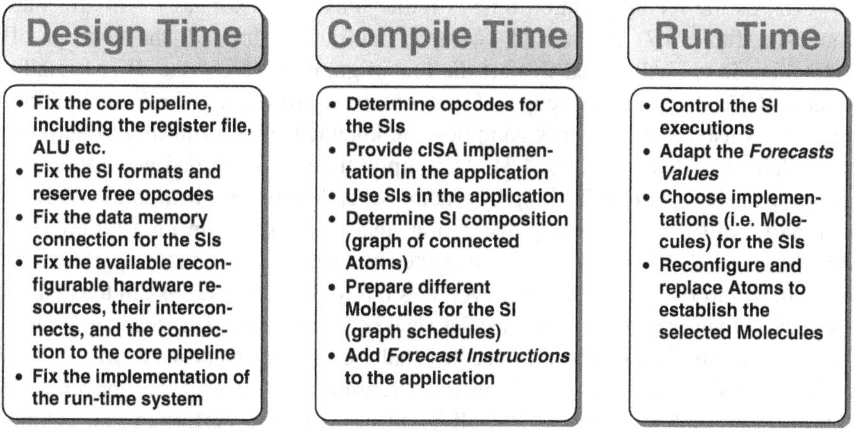

Design Time	Compile Time	Run Time
• Fix the core pipeline, including the register file, ALU etc. • Fix the SI formats and reserve free opcodes • Fix the data memory connection for the SIs • Fix the available reconfigurable hardware resources, their interconnects, and the connection to the core pipeline • Fix the implementation of the run-time system	• Determine opcodes for the SIs • Provide cISA implementation in the application • Use SIs in the application • Determine SI composition (graph of connected Atoms) • Prepare different Molecules for the SI (graph schedules) • Add *Forecast Instructions* to the application	• Control the SI executions • Adapt the *Forecasts Values* • Choose implementations (i.e. Molecules) for the SIs • Reconfigure and replace Atoms to establish the selected Molecules

Fig. 4.2 Fix at design/compile time and adapt at run time

---

[1]For instance, depending on the number of ports of the register file.

[2]It does not need to be determining yet which opcode shall correspond to which SI; just opcodes for SIs need to be reserved.

represent the data flow) to define the SI functionality. In addition, the SI needs to be used in the application[3] and the assembler needs to be extended[4] to obtain the binary. The SI graph can be used to prepare the different molecules for the SI. Preparing the molecules corresponds to applying different resource constraints to the SI graph and then scheduling the execution of the atoms accordingly (as shown in Fig. 3.4). Actually, the molecule information is not demanded at compile time and it could also be created at run time. However, it is possible to prepare the molecules at compile time because they do not change during run time (note: which molecule is used to implement an SI typically changes during run time, but the individual molecule schedules are not affected). As the amount of information demanded to store the individual molecule schedules is relatively small, molecule schedules can be prepared during compile time. In contrary to the manually designed SIs and atoms (see Sect. 3.3), a tool was developed in the scope of that work presented in this monograph that automatically creates all molecules for a given SI graph, because the number of molecules is rather large in comparison to the number of SIs and atoms, and thus a manual molecule creation would be very time consuming. The resulting molecule schedules are saved and the information is provided to the run-time system. This corresponds to a particular trade-off between run-time computation effort and run-time storage requirements that may be changed, depending on the constraints of the target system.

One further step needs to be done at compile time: the run-time system needs to be triggered to start the reconfigurations. As a run-time reconfigurable fabric is used to provide SI implementations on demand, some information must be used to trigger the run-time system to start these reconfigurations. One possibility would be to wait until an SI shall execute and to start the corresponding reconfiguration then. However, as the reconfiguration time is rather long (approximately 1 ms for one atom, see Table 5.7) this would be rather inefficient. Instead, the concept of prefetching is typically used to start the reconfiguration before the SI is actually demanded [LH02]. In the scope of the presented work, a hybrid prefetching approach is used that combines compile-time knowledge with run-time adaptivity (as explained in detail in Sect. 4.3). At compile time, it is decided when the run-time system shall start prefetching by placing so-called forecast instructions (FI) in the binary. An FI is an assembly instructions that informs the run-time system that one or multiple SIs are expected to be executed in the next time (and thus prefetching for them would be beneficial). However, it is left to the run-time system to eventually decide whether or not some molecules shall be reconfigured (i.e., their composing atoms) for these SIs and which molecules shall be used for a particular SI.

The compile-time task focuses on (a) determining a place in the applications control-flow graph (where the FI shall be placed) and (b) performing an offline

---

[3]For example, by calling the SI using inline assembly or by modifying the compiler to automatically use it.

[4]That is, the tool that creates binary code out of assembly code needs to know which instruction format and opcode shall be used for an SI that occurs in the assembly code.

profiling to predict how often an SI is expected to execute (the so-called *forecast value* of an FI). On the one hand, the FI should execute early enough to provide sufficient time to complete the reconfigurations (in best case). On the other hand, if it is *too early* then the chance increases that between the FI and the targeted SIs other SIs demand the reconfigurable fabric and their hardware is not loaded or even replaced by that *too early* FI. For instance, when considering the H.264 video encoder flow from Fig. 3.5 the SI for the in-loop de-blocking filter (LF) could be forecasted during the execution of the motion estimation (ME). Certainly, that would provide sufficient time to reconfigure the SIs for LF. However, as the encoding engine (EE) executes between ME and LF, the SI for LF would allocate atom containers (ACs) on the reconfigurable fabric without providing actual benefit for the upcoming SI executions (i.e., for the SIs from EE), that means the FI for LF was *too early*. For the application flow of the H.264 video encoder, the FIs were placed between the three major computational blocks ME, EE, and LF and performed a compile-time profiling to determine the corresponding forecast values that predict the expected SI execution frequency. Placing the FIs inside the computational blocks before the actual SI executions would lead to frequent reconfigurations and replacements within each computational block. This is not beneficial due to the relatively long reconfiguration time of one atom (approximately 1 ms per atom, see Table 5.7) and the relatively short execution time (30 frames per second corresponds to 33 ms per frame, i.e., sufficient time to reconfigure approximately 33 v per frame).

After the design-time and compile-time decisions determined certain architecture settings and prepared the application, the task for the run-time system is to provide adaptivity considering the SI implementation alternatives and their reconfigurations. Recalling the SI execution frequency analysis in Fig. 3.6, it may depend on input data which SI is demanded how often. Therefore, a compile-time profiling (to determine the forecast values) is not sufficient. Instead, the forecast value needs to be updated at run time to reflect recent SI execution frequencies. An online monitoring and an online prediction scheme is used to update the expected SI execution frequencies dynamically. Figure 4.3 provides an overview of the run-time system, showing the monitoring and prediction. The *decoder* of the run-time system observes the instruction stream and triggers subsequent modules of the run-time system when detecting SIs or FIs. When an SI is observed, the *execution control* manages how it shall be handled (using the cISA or a hardware molecule that uses the atom infrastructure). The *online monitoring* counts the number of executed SIs [BSTH07]. Later, the *prediction* determines the difference between the initial FI predictions (i.e., the forecast values) and compares it with the observed SI executions from the monitoring (details are explained in Sect. 4.3). This difference is used to update the forecast value for the next execution (in the H.264 example this means for the next video frame) [BSTH07].

The updated SI execution frequency prediction is used in the *selection* to determine which molecules shall be loaded (details are explained in Sect. 4.4) [BSH08c]. In addition to the execution frequency, the amount of available ACs is used as input for the molecule selection, as the availability of reconfigurable fabric may change during run time (depending on changing number and priorities of tasks), as

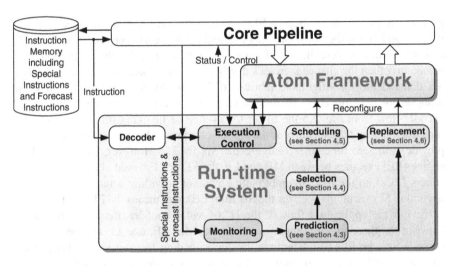

Fig. 4.3 Overview of the RISPP run-time system

motivated in Sect. 3.2. Depending on the selected molecules, certain atoms need to
be reconfigured. As the atom infrastructure only provides one reconfiguration port
(i.e., only one atom after the other may be reconfigured; not multiple at the same
time), the *scheduling* determines the sequence in which the atoms shall be loaded
to the reconfigurable fabric (details are explained in Sect. 4.5) [BSKH08]. As the
molecule selection is performed during run time, the atom reconfiguration scheduling cannot be prepared at compile time and needs to be executed at run time as well.
Whenever a new atom shall be reconfigured and the atom infrastructure is full (i.e.,
no atom container is available), the *replacement* decides which atom shall be
replaced to load the new atom (details are explained in Sect. 4.6) [BSH09b].

SI execution frequency prediction, molecule selection, atom reconfiguration
scheduling, and atom replacement are the main algorithms of RISPP's run-time
system. Before explaining them in detail in the subsequent sections, this section
presents an overview in Fig. 4.4 and discusses how these algorithms interact with
each other using state transitions [BSH08b]. The initial state (Decode) triggers the
execution of substate machines that are executed in parallel with the decode state.
Note that a state does not necessarily correspond to a single clock cycle. Some
subsequent states are executed in the same cycle (or pipelined), whereas other states
require multiple cycles. SIs or FIs trigger the components of the run-time system,
as shown in Fig. 4.4. On the implementation side, the run-time system can be partitioned into two parts: A *synchronous part* running in the clock domain of the core
pipeline and an *asynchronous part* that may run in a different clock domain
(although the core pipeline triggers its execution). The synchronous part comprises
fine-tuning the forecast values (lower left box) and all parts that are triggered by SIs
(upper boxes). These parts have to be synchronous to the core pipeline, as they are
tightly coupled to it (e.g., sending control signals or receiving status information).
While adding the synchronous part to RISPP's hardware prototype, high effort

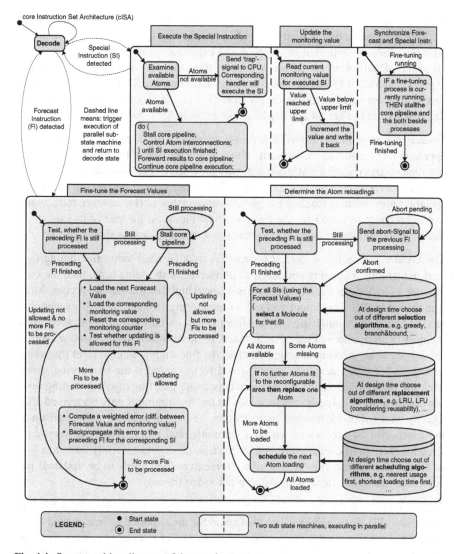

Fig. 4.4 State-transition diagram of the run-time system

was spent to make sure that the critical path of the core pipeline is not affected. The asynchronous part of the run-time system comprises all atom reloading decisions (see lower right box).

Execute the special instruction. When an SI shall execute, it depends on the available atoms, which molecule will be used to implement it. As this condition changes relatively seldom,[5] a small memory stores for each SI the information which molecule

---

[5]At least one atom needs to be loaded which lasts approximately 1 ms.

shall be used to implement it. If insufficient atoms for a hardware implementation are available, then a trap is send to the CPU and the trap handler will execute the SI with the cISA as motivated in Chap. 3 and explained in more detail in Sect. 5.2. However, if sufficient atoms are available, then the core pipeline is stalled (as explained in Sect. 4.1) and the atoms are used to execute the SI. Therefore, the atoms need to be connected correspondingly to realize the SI execution, as motivated by the SI schedule in Fig. 3.4 and explained in detail in Sect. 5.4.

Update the monitoring value. The run-time system uses an online monitoring approach for fine-tuning the forecast values, which are one of the main inputs to the run-time system (besides #ACs and statically provided information about SIs, atoms, etc). Whenever an SI executes, an SI-specific counter is increased until a hardware-specific upper limit is reached (to prevent an overflow). This counter reflects the execution count of the SI for the computational block. At the end of its execution, these counter values are used to fine-tune the forecast values, as shown below.

Synchronize forecast and special instruction. If a forecast value fine-tuning (see next paragraph) is running and an SI shall execute at the same time, then the core pipeline is stalled and the SI execution is delayed until the fine-tuning is completed. The reason is that fine-tuning and online monitoring (triggered by SI executions) read and write the monitoring data and inconsistencies may occur if both access this memory in an interleaved way. For instance, if the monitoring increments a counter value (i.e., read, increment, and write) and the fine-tuning aims to reset that value to zero in between, then this "reset" would be discarded if the monitoring writes back the old (incremented) value afterwards. Note that this pipeline stalling does not lead to a noticeable performance loss, as the fine-tuning is only performed after a computational block finished its execution, i.e., rather seldom in comparison to the number of SI executions within a computational block. Even the worst-case situation (i.e., an SI execution right after the fine-tuning started) only leads to a loss of few cycles (depending on how many forecast values needs to be updated) per computational block (and therefore also per video frame).

Fine-tune the forecast values. The process of fine-tuning is triggered by a forecast instruction (FI). Such an FI predicts that some SIs are expected to be executed soon. In addition, the FI is used to end previous predictions. For instance, when the encoding engine in the H.264 video encoder is about to execute, then an FI predicts the SIs for EE and at the same time informs, that the SI for the motion estimation (ME, i.e., the previously executed computational block, see Fig. 3.5) are no longer demanded. When such an *ending* FI information for an SI, e.g., SAD for ME, is reached then fine-tuning its forecast value starts. The difference between the previous forecast value of SAD (used in the FI before ME) and its monitoring value (i.e., the actual execution count) is used to calculate a modified forecast value for the FI that is placed before ME. The details about different scaling factors to weight and back propagate the error are examined in Sect. 4.3. In the case that a fine-tuning operation is still processing when an *ending* FI is reached (which would start another fine-tuning), the process needs to wait until the preceding fine-tuning is finished. The reason is that the counter values from online monitoring would not be reset for the not yet processed forecast values, which then would lead to wrong counter values in the next loop iteration. The same is true if an SI executes while a fine-tuning operation is currently running (see "synchronize forecast and special instruction" in Fig. 4.4 and in the previous paragraph).

Determine the atom reloadings. The asynchronous part of the run-time system comprises the atom reloading decisions. Although this part is triggered by an FI, it can execute independently after it was started. In contrast to fine-tuning the forecast values, a previously started execution of this part may be aborted, as the SIs of the preceding FI are generally obsolete when a new one arrives (otherwise, they will be predicted again in the new FI). The forecast values are used together with (at compile-time determined) information of the different SI implementations to select implementations for all requested SIs. As this decision has to be made at run time, the computational overhead is critical and therefore, a greedy heuristic is deployed (see Sect. 4.4). This heuristic iterates over all molecules of the forecasted SIs and determines a profit value for each one. The locally best molecule is selected and the same procedure is repeated for the remaining SIs. As soon as the first reconfiguration is started (after the first molecule is selected), more time may be spent in determining further decisions, as it is then done in parallel to the reconfiguration. For the selected molecules, the atoms need to be loaded sequentially.[6] Determining a specific reconfiguration sequence may have a high impact on the application execution time. For instance, multiple SIs might be required for the upcoming computational block and some of them might be executed significantly more often than others. In this case, it might be beneficial to reconfigure the atoms for the more often required SIs at first, depending on the performance differences of the corresponding molecules. In the scope of the presented work, an approach is used that locally selects those atoms that lead to the locally best performance improvement due to a better molecule (see Sect. 4.5). Finally, some currently available atoms may need to be replaced to offer new molecules. If nearly all atoms are going to be replaced, then the decision which atoms shall be kept has only a small performance impact. However, in situations where only relative few atoms are going to be replaced (e.g., due to insufficient reconfiguration time), it is beneficial to keep those atoms that will be required again soon. As the upcoming requirements are not known beforehand, those atoms are replaced that lead to locally smallest performance degradation of all required SIs of the application (see Sect. 4.6).

4.2.1 Summary of the Requirement Analysis and Overview

In this section, it was analyzed which degree of flexibility is needed in the architecture to exploit the adaptivity and performance that is offered by the novel concept of modular special instructions (SIs). The decisions were partitioned into three time

---

[6]Typically only one reconfiguration port is available.

domains, i.e., design time, compile time, and run time. Afterwards, this section discussed which decisions can be fixed at design time (e.g., defining the data memory bandwidth, size of reconfigurable fabric, SI interface, etc.), which can be fixed at compile time (e.g., defining SIs/molecules/atoms, using the forecast instructions, etc.), and which decision need to be left to the run-time system (e.g., deciding which molecule shall be used to implement an SI etc.). Then, a first over-view of the novel run-time system was given, showing which components it demands and how they are connected. Afterwards, the tasks of these components and their interactions were presented and in the following sections, the developed solutions for these components, component-specific analysis, algorithmic alterna-tives, and implementation results are presented.

4.3 Online Monitoring and Special Instruction Forecasting

The concept of modular SIs allows us to decide dynamically which molecule shall be used to implement an SI. This allows us to distribute the available reconfigurable fabric (i.e., the number of atom containers) to the SIs, depending on the application requirements that may change during run time (as motivated in Sect. 3.3). However, to be able to determine a *good* distribution of the reconfigurable fabric (i.e., reflect-ing the application requirements), the information "how often" each SI is demanded is needed. An offline profiling is used to obtain this information for a particular application run. As the SI execution frequency may change at run time, online monitoring and a so-called fine-tuning is used to reflect these changes and to fore-cast the expected SI execution frequencies [BSTH07].

Forecast value (FV). A FV is a number that predicts the expected execution fre-quency of a particular SI, i.e., an FV is specific for an SI. The prediction is placed into the application binary using forecast blocks and forecast instructions (see below). During application run time, the value of the prediction may be changed; however, the place in the binary is fixed.

Forecast block (FB). A FB is a set of predictions, i.e., multiple SIs are predicted with individual FVs at the same time. Combining multiple forecasts to one block is important, as the occurrence of a forecast triggers the run-time system to perform the corresponding reconfigurations. For instance, multiple forecasts would be treated as individual events, then the last forecast would invalidate the previous ones and thus only the last forecast would be considered by the run-time system (see also Fig. 4.4).

Forecast instruction (FI). A FI is an assembly instruction that is used to notify the occurrence of an FB to the run-time system. A FB may be realized by multiple FI calls, where each FI forecasts a particular SI (i.e., provides the SI opcode and the FV). However, for the implementation, an FB is realized as a single FI that points to a dedicated forecast memory (using starting address and length) that contains all forecast information. This approach comes with the advantages that less instructions are needed in the application binary and the pipeline (fewer cycles needed to execute an FB) and that – to be able to modify the FVs – it does not depend on writable instruction memory (i.e., read-only instruction memory is sufficient). In addition, providing the forecast information (i.e., SI opcode and FV) is independent of the length of the instruction words (typically 32-bit instructions). Only a new assembly instruction needs to be developed that provides a start address and length in the forecast memory (see Sect. 5.1 for the instruction format), but the forecast memory may contain more data than would fit into a 32-bit instruction word. For instance, in addition to the FV, the molecules selection (see Sect. 4.4) demands additional information that is provided during offline profiling but not updated during run time.

Figure 4.5 shows a typical application scenario that allows us to explain the fundamental idea of fine-tuning the FVs in Fig. 4.6. The example contains two forecast blocks FB_1 and FB_2 (note, an FB does not necessarily have to be an individual base block, as – for simplicity – indicated in the figure). FB_1 predicts the

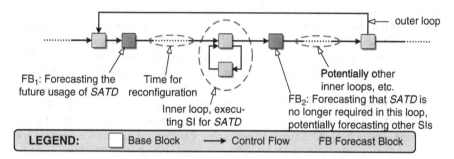

Fig. 4.5 Example control-flow graph showing forecasts and the corresponding special instruction executions

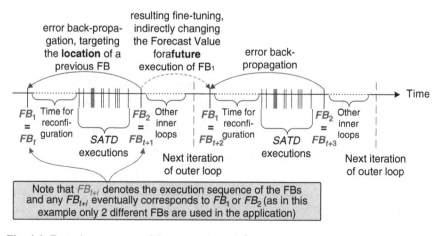

Fig. 4.6 Execution sequence of forecast and special instructions with the resulting error back propagation and fine-tuning

execution of an SI of an execution kernel (inner loop) and FB_2 informs that the kernel finished execution and the SI is no longer demanded. In this example, the SI corresponds to SATD that was introduced in Fig. 3.3, and the execution kernel could represent the motion estimation (actually, it would additionally demand the SAD SI). The "potentially other inner loops" after FB_2 could correspondingly represent the encoding engine and the in-loop de-blocking filter of the H.264 video encoder, as introduced in Fig. 3.5, and the outer loop would iterate over the video frames. This shows how the actual SI execution is encapsulated into two FBs.

Figure 4.6 presents a time axis that shows the execution flow of the control-flow graph in Fig. 4.5 for two iterations of the outer loop. It becomes apparent how the SATD executions are encapsulated between FB_1 and FB_2. Whenever FB_2 is reached, the initial expectation from FB_1 (i.e., the prediction how often SATD will be executed) is compared with the actual execution count from the online monitoring. The resulting error is back propagated, i.e., the initial expectation from FB_1 is fine-tuned. A technique that uses differences between distinct points in time is called temporal difference (TD) scheme [Sut88] and in the scope of the work that is presented in this monograph, it is used for SI forecasting. The TD scheme is based on the Markov property, this means that the conditional property to reach a specific FB and to count a specific number of SI executions between the last FB and the reached FB must only depend on the previous FB and not on the chain of preceding FBs. Sutton and Barto explain that this Markov property cannot be guaranteed in real-world problems, but that experiments suggest, that the TD scheme nevertheless achieves good results in practice [SB98]. Therefore, the Markov property is mainly used to derive the TD scheme and to prove its convergence. However, convergence is less important for the presented approach, as the SI execution frequency varies and thus convergence is not the target but adaptation is (as also noted by Sherwood et al. [SSC03]).

The example from Fig. 4.6 demonstrated that any control-flow graph can eventually be represented as a linear chain of FBs that are connected in their execution sequence. Figure 4.7 shows such a chain of FBs. Please note that the FBs are

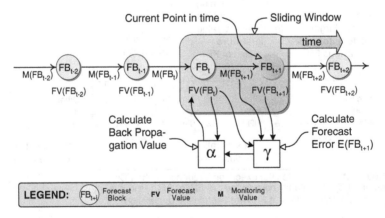

Fig. 4.7 A chain of forecast blocks, showing how the information in a sliding window is used to determine a forecast error that is back propagated

labeled according to their execution time: FB_{t+1} is executed right after FB_t, i.e., there are no other FBs in between but potentially many SIs and instructions from the cISA that are omitted for clarity. The representation as linear execution sequence implies that any two different forecast blocks FB_x, FB_y, $x \neq y$ in this chain may eventually correspond to the same FB in the application binary, just at different points in time during the application execution. For instance, in Fig. 4.6 FB_t and FB_{t+2} correspond to FB_1, whereas FB_{t+1} and FB_{t+3} correspond to FB_2. As an extreme case, the application may only contain one FB in its binary that is executed several times and thus all entries in the chain correspond to the same FB. However, for fine-tuning the FVs, only the sequence of FB executions needs to be known, independent of their mapping to point in the application binary. In particular, only the directly preceding forecast block FB_t needs to be known when reaching FB_{t+1} to be able to update the FV at FB_t. This is indicated by the sliding window in Fig. 4.7 and is explained in the following.

4.3.1 Fine-Tuning the Forecast Values

The processes of fine-tuning the forecast values at FB_t when reaching FB_{t+1} for a particular SI is explained in detail, using the illustration in Fig. 4.7. This means that all FBs in this figure correspond to forecasts for that particular SI.[7] Note that the same description holds for any SI and fine-tuning the FVs for multiple SIs is done independent of each other. The monitoring counts the executions of that SI between two subsequent forecast blocks FB_t and FB_{t+1} as $M(FB_{t+1})$. In the hardware implementation, this corresponds to a memory location that is reset to zero after the fine-tuning process at FB_{t+1} is completed (to count the subsequent SI executions in the same memory location afterwards). This monitoring value is compared with the FV from FB_t. The difference between $FV(FB_t)$ and the monitoring value $M(FB_{t+1})$ corresponds to an error of the FV, i.e., FB_t predicted a certain amount of SI executions but $M(FB_{t+1})$ executions were monitored. In general, the forecast block FB_{t+1} may also predict some executions of that SI, i.e., FB_t predicted a certain amount of SI executions, $M(FB_{t+1})$ SI executions were monitored so far, and $FV(FB_{t+1})$ executions are predicted to come soon. Therefore, the FV of FB_{t+1} needs to be considered in addition to $FV(FB_t)$ to calculate the error $E(FB_{t+1})$.

$$E\left(FB_{t+1}\right) := M\left(FB_{t+1}\right) - FV\left(FB_t\right) + \gamma\, FV\left(FB_{t+1}\right) \qquad (4.1)$$

$$FV\left(FB_t\right) := FV\left(FB_t\right) + \alpha\, E\left(FB_{t+1}\right) \qquad (4.2)$$

Equation 4.1 shows how the error is calculated. The parameter $\gamma \in [0,1]$ is used to adjust how strong the FV of FB_{t+1} should contribute to the error. A relatively large

---

[7]Note: in general, the FBs may additionally contain forecasts for further SIs.

value for γ would consider the prediction of FB_{t+1} to a large degree. This is potentially problematic for the following two reasons:

1. The prediction of FB_{t+1} is used to calculate the error, i.e., the error is no longer actually observed, but it is mixed with expectations about the future. If these expectations are not correct, then the quality of the calculated error is affected in a negative way.
2. The calculated error is used to fine-tune the previous forecast. Considering the prediction of FB_{t+1} to update the FVs of earlier FBs will eventually shift the prediction of FB_{t+1} to an earlier point in time (i.e., to FB_t). In the next iteration of the loop, this information may be shifted backwards to FB_{t-1} and so on. As motivated in Sect. 4.1, predicting an SI execution *too early* may lead to a reduced performance, as the run-time system is advised to provide hardware accelerators for that SI, although it is not needed yet. This affects the availability of reconfigurable hardware for SIs that are actually demanded.

After calculating the error, it is back propagated to the preceding FB. The strength of this back propagation is adjusted with the parameter $\alpha \in [0,1]$, see Eq. 4.2. This parameter allows to avoid *thrashing* (i.e., rapid alternation between two extreme values). For instance, consider the following situation: if the execution of an SI in a computational block alternates between 0 and 100 executions for subsequent executions of that block (i.e., 0, 100, 0, 100, ...), the computational block is encapsulated by FB_1 and FB_2 (like in Fig. 4.5), $FV(FB_1)$ is initialized with 100, $FV(FB_2)$ is constant 0, and α and γ are set to 1 and 0, respectively, then the fine-tunings of FB_1 will occur as shown in Example 4.1.

Example 4.1 Thrashing of forecast value due to nonweighted error back propagation.

    1st fine-tuning: old $FV(FB_1) = 100, M(FB_2) = 0,$
                             calculated error $E(FB_2) = 0{-}100 = -100,$
                             $\rightarrow$ new $FV(FB_1) = 100 + 1 \times (-100) = 0.$
    2nd fine-tuning: old $FV(FB_1) = 0, M(FB_2) = 100,$
                             calculated error $E(FB_2) = 100{-}0 = 100,$
                             $\rightarrow$ new $FV(FB_1) = 0 + 1 \times (100) = 100.$
    3rd fine-tuning: old $FV(FB_1) = 100, M(FB_2) = 0,$
                             calculated error $E(FB_2) = 0{-}100 = -100$
                             $\rightarrow$ new $FV(FB_1) = 100 + 1 \times (-100) = 0.$
    4th fine-tuning: ...

This shows that the monitoring value may completely overwrite the previous forecasts, and in this particular situation, the forecast value is always the actual opposite of the actual SI execution frequency. Please note that $FV(FB_2) = 0$ is constant, because the SI is not executed between FB_2 and FB_1, therefore the monitoring value is $M(FB_1) = 0$, therefore the calculated error is $E(FB_1) = 0$, and therefore the initial FV stays unchanged. To avoid the observed *thrashing* in the FV, a more moderate value of, e.g., $\alpha = 0.5$ can be used which would lead to the fine-tunings shown in Example 4.2.

Example 4.2 Avoiding forecast thrashing by weighting the error back propagation.

1st fine-tuning: old $FV(FB_1) = 100$, $M(FB_2) = 0$,
                         calculated error $E(FB_2) = 0–100 = -100$
                         $\rightarrow$ new $FV(FB_1) = 100 + 0.5 \times (-100) = 50$.
2nd fine-tuning: old $FV(FB_1) = 50$, $M(FB_2) = 100$,
                         calculated error $E(FB_2) = 100–50 = 50$
                         $\rightarrow$ new $FV(FB_1) = 50 + 0.5 \times (50) = 75$.
3rd fine-tuning: old $FV(FB_1) = 75$, $M(FB_2) = 0$,
                         calculated error $E(FB_2) = 0–75 = -75$
                         $\rightarrow$ new $FV(FB_1) = 75 + 0.5 \times (-75) = 37.5$.
4th fine-tuning: old $FV(FB_1) = 37.5$, $M(FB_2) = 100$,
                         calculated error $E(FB_2) = 100–37.5 = 62.5$
                         $\rightarrow$ new $FV(FB_1) = 37.5 + 0.5 \times (62.5) = 68.75$.
5th fine-tuning: old $FV(FB_1) = 68.75$, $M(FB_2) = 0$,
                         calculated error $E(FB_2) = 0–68.75 = -68.75$
                         $\rightarrow$ new $FV(FB_1) = 68.75 + 0.5 \times (-68.75) = 34.375$.
6th fine-tuning: …

It is apparent, that the prediction is still not correct (which is not surprising, considering the extreme alternation); however, with a moderate α value, the changes in the prediction are less rapid and most notably the accumulated absolute error is smaller (i.e., the FV was closer to reality). In the extreme case of $\alpha = 0$, the calculated error would not be back propagated at all, which means that the initial FV (determined from offline profiling) would be constant. This shows that the so-called static prefetching (i.e., the values are not updated during run time [LH02]) is a special case of this model.

$$\forall i \geq 0 : FV\left(FB_{t-i}\right) := FV\left(FB_{t-1}\right) + (1-\lambda)\lambda^i \alpha E\left(FB_{t+1}\right) \tag{4.3}$$

$$\lambda = 0 \rightarrow (1-\lambda)\lambda^i = (1-0)0^i = \begin{cases} 1, i = 0 \\ 0, i > 0 \end{cases} \tag{4.4}$$

In addition to back propagating the calculated error to the directly preceding FB, it is also possible to propagate it back to multiple FBs. This can potentially increase the adaptation rate, as FBs that are not predecessors for a particular FB can be affected by the monitored SI executions in a direct way. If only the directly preceding FBs are updated, a modified information will also propagate back to earlier FBs; however, it demands multiple iterations of an outer loop until this eventually happens. Figure 4.8 visualizes this back propagation to potentially all previously executed FBs. The strength of the back propagation to the farther away FBs is diminished by the parameter $\lambda \in [0,1]$ to assure that the prediction of the monitored SI execution count is used in a rather local scale. The factors shown in the figure are multiplied with the weighted error as it was calculated in Eq. 4.2, altogether leading to the back propagation

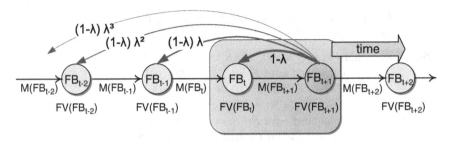

Fig. 4.8 A chain of forecast blocks, showing how multiple previous forecast blocks may be updated, depending on parameter λ

shown in Eq. 4.3. In the extreme case of $\lambda = 0$, the back propagation of Eq. 4.3 corresponds to the one in Eq. 4.2 [i.e., Eq. 4.2 is a special case], because for FB_i (i.e., $i = 0$) the new factor $(1 - \lambda)\lambda^i$ evaluates to 1, whereas for all other FBs (i.e., $i > 0$) it evaluates to 0 [see Eq. 4.4] and thus the FV of these FB remains unchanged.

When considering a hardware implementation for the temporal difference (TD) scheme, allowing $\lambda > 0$ implies a significant overhead. Relatively large amounts of FBs have to be traced for a particular SI as they will be updated when a new FB for that SI is reached. It has to be noted, that the sequence of the FB executions is important for the updates, as visible in Fig. 4.8. Therefore, the history of FB executions that have to be traced for potential fine-tunings is only bound by the application execution time, but not by the statically known number of FBs in the application binary. However, for an implementation the number of maximally traceable FBs for a particular SI is actually bound by the amount of memory that is provided for storing that execution sequence. In addition, the number of preceding FBs that have to be fine-tuned also determines the processing time for the fine-tuning process. Therefore, only an approximation for the TD scheme can be implemented in hardware. If the parameter λ would be fixed to zero, then the hardware implementation cost and algorithm execution time for the TD scheme would reduce significantly. For each SI only the last FB that forecasted this SI needs to be memorized and fine-tuned. Please note, that this would not violate the general TD scheme, which allows to choose the parameter λ from [0,1]. However, it affects the behavior of the fine-tuning process. In particular, it may lead to a *slower* back propagation. Consider a code region that executes FB_1, FB_2, and FB_3 sequentially, all of them forecasting a particular SI. For $\lambda = 0$, a forecast error that is computed for FB_3 would only be back propagated to the directly preceding FB_2, but FB_1 would not be affected by this error. At a later execution of the same three FBs, the error that was already back propagated to FB_2 would now (to some degree, i.e., in a weighted form) be back propagated to FB_1. Therefore, it takes two iterations to move information from FB_3 to FB_1. For $\lambda > 0$, a forecast error at FB_3 would directly affect FB_1, i.e., the information is back propagated in a faster way. In the simulation environment, the TD scheme is implemented with adjustable parameters for α, γ, and λ to evaluate the performance impact for the application execution time (due to different FV fine-tuning and the corresponding reconfigurations) when restricting λ to zero. For $\lambda = 0.6$ an application speedup between 1.01× and 1.10× is observed in comparison to $\lambda = 0$. Averaged over different application sequences and different sizes of reconfigurable fabric, an application speedup of 1.05× is obtained. The peak performance improvement of 1.44× was observed for $\lambda = 1$; however, in a different application sequence the performance degraded to 0.91× and in average 1.08× is observed in comparison to $\lambda = 0$. As an average application speedup of less than 10% does not justify significantly increased hardware requirements, the value for λ is restricted to zero in the hardware prototype.

As the parameters α and γ do not affect the hardware requirements significantly[8] both parameters are provided in an adjustable manner (i.e., they can be configured by the application) in the prototype instead of hardwiring them. In addition, the feature to explicitly forbid fine-tunings for particular FVs is provided (as indicated by the "updating allowed" control edge in Fig. 4.4. This can be beneficial if the application programmer knows the application well and wants to avoid potentially misleading fine-tuning operations on selected FVs. Typically, the FBs that inform that SIs are no longer demanded are declared to be nonupdatable, to assure that the FVs stay zero. The reason is that the application programmer knows that, for instance, the time between the end of the motion estimation (ME) and the start of the ME for the next frame is relatively long. Therefore, it is not beneficial to reserve parts of the reconfigurable fabric for these SIs. To assure this, an FV of zero is demanded for the ME SIs after ME is completed. However, depending on the parameters α and γ, the FVs in the FB that indicates the end of the ME may gradually increase. When the FB right before ME is reached, then the FB right after ME (of the previous frame) will be fine-tuned (because all other FBs in between target different SIs). The FB after ME predicted zero executions and also zero executions were monitored since then. However, depending on the value of γ, the prediction of the FB before ME will be considered for calculating the error and depending on the value of α, this error will be back propagated, leading to an FV that is larger than zero. To avoid this, either γ can be configured to zero or the FB after ME can be declared to be nonupdatable (both possibilities are provided in the system).

4.3.2   Evaluation of Forecast Fine-Tuning

The impact of the parameters α and γ is evaluated before looking into the actual forecast adaptation details. Figure 4.9 shows the impact of different parameter values on the overall application execution time. It becomes noticeable that the execution time is dominated by the parameter α. Setting α to zero corresponds to static prefetching, i.e., the calculated error is not back propagated (see Eq. 4.2) and $\alpha = 1$ corresponds to a rapid adaptation to varying situations. This leads on average (over the γ values) to a 1.24× faster execution time than using $\alpha = 0$.

The changes of the FVs are analyzed in detail, i.e., in addition to presenting the application execution time for different settings for α, the changes of FVs over time

---

[8]Both demand a multiplication.

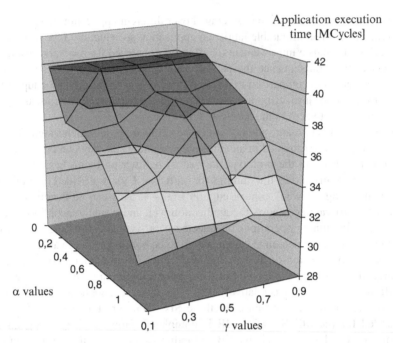

Fig. 4.9 Parameter evaluation for α and γ for $\lambda = 0$, showing the resulting application run time

are examined. This will unveil how they adapt to the monitored SI execution frequencies, which eventually affects the overall application execution time, as the FVs are used as main input for the molecule selection (see Fig. 4.4). For benchmarking, the video sequence was applied that was already used in Sect. 3.3 to motivate the need for adaptation when facing changing ratios of I-MBs and P-MBs (see Fig. 3.6). Figure 4.10 presents an excerpt of this video sequence (frames 150–525 where the adaptation is especially visible), showing the actually demanded number of I-MBs in comparison with the predicted number of I-MBs for different values of α. Please note that in this particular example the value of γ does not change the results. The reason is that the application only contains two FBs (FB_{start} and FB_{end}) for the SIs that encode I-MB and P-MBs, FB_{start} predicts their occurrence with the corresponding FVs and FB_{end} denotes that these SIs are no longer used in this frame. To assure that the FVs of FB_{end} stay zero, it is declared as "not updatable," as described above. Therefore, when back propagating from FB_{start} to FB_{end} (executed in the previous frame), the value of FB_{end} is not affected (independent of γ) and when back propagating from FB_{end} to FB_{start}, the FV of FB_{end} (which is weighted by γ) is always zero.

The effect of the different α values becomes noticeable in Fig. 4.10. At first, it can be noticed that all different forecasts follow the curve of actual SI executions. However, they all face a certain offset, i.e., the predicted curve appears to be shifted to the right (on the time axis) in comparison to the actual executions. This is because the back propagation requires some time until it adapts to a changed situation. Here, a smaller value of α leads to a distinctive averaging/smoothing of the curve, i.e., the high peaks of the actual SI executions are not reflected in the predicted curve

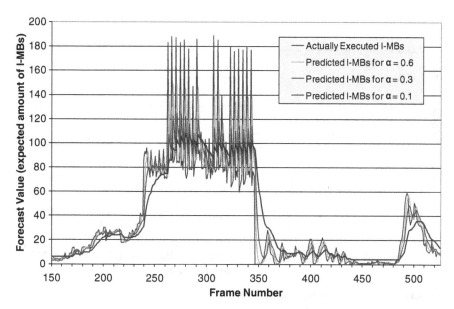

Fig. 4.10 Evaluation of the forecast value for different values of α, showing the actual and predicted SI execution

(as motivated in the *thrashing* example above). This averaging offers advantages during rapidly changing scenes. This becomes noticeable when analyzing the predicted curve for $\alpha = 0.6$ in the hectic scene between frames 260 and 340. Whenever the actual execution faces a high peak of I-MBs, the curve follows that peak shortly afterwards. However, at the time when the curve actually follows, the peak already ended and thus the prediction is wrong again. This corresponds to the initially motivated example of *thrashing*. In the same scene, the curve for $\alpha = 0.1$ only follows the actual SI executions to some degree. It raises the expected amount of I-MBs to a noticeably higher level than it was before, but it does not follow individual peaks and therefore leads to a better prediction. However, at larger changes of the average amount of I-MBs (i.e., when the offset around which the fluctuations appear changes) – for instance, when entering the hectic scene (around frame 250) or when leaving it (around frame 350) – the curve for $\alpha = 0.1$ is hardly able to follow that change. Here, the prediction error is noticeably bigger than for larger α values.

In Fig. 4.11 the accumulated absolute forecast error is investigated further, i.e., for each frame the difference between the initially predicted number of I-MBs and the actually executed number of I-MBs is calculated and then the absolute values of these errors are accumulated.[9] For each frame, the figure shows the accumulated absolute error since the first frame. At the beginning, the curve for $\alpha = 0.1$ leads to the worst accumulated absolute error until a local maximum at frame 250. In this beginning, the stronger adaptation leads to a noticeably better prediction. However, as soon as

---

[9]The fine-tuning actually works on the number of SI executions and not on the number of MBs that are shown here for simplicity.

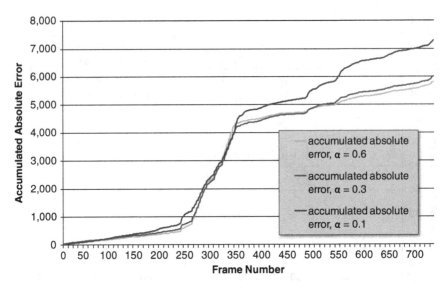

Fig. 4.11 Accumulated absolute forecast error

the hectic scene starts, the curve for $\alpha = 0.1$ performs noticeably better and is even able to close the initial gap. Actually, between frames 310 and 340 $\alpha = 0.1$ even leads to a smaller accumulated absolute error than $\alpha = 0.6$ (coming close to $\alpha = 0.3$), i.e., the initially worse performance was more than compensated. Actually, $\alpha = 0.6$ appears to be the worst strategy during that time, giving away the initially gained advantage. However, right after the hectic scene ends (around frame 350), $\alpha = 0.1$ immediately looses its advantage and becomes the worst overall strategy again (with a widening gap in comparison to the other parameter settings). From frame 489 onwards, $\alpha = 0.6$ again leads to the best (i.e., smallest) accumulated absolute error in comparison to the other strategies. This shows that $\alpha = 0.1$ only provides advantages in certain rapidly changing situations, whereas $\alpha = 0.6$ may lead to the best overall forecast and $\alpha = 0.3$ corresponds to a kind of compromise. Therefore, the parameters are kept flexible in the hardware prototype, i.e., a so-called helper instruction is offered (see Sect. 5.1) that allows the application to set and change these parameters.

4.3.3 Hardware Implementation for Fine-Tuning the Forecast Values

For the prototype, a hardware implementation for the presented forecasting scheme is implemented. Figure 4.12 shows the pipelined implementation of it, which is explained in the following. The instruction decoder triggers the forecasting unit when a forecast instruction (FI, indicating an FB) is executed. Afterwards, the forecast unit runs in parallel to the processor pipeline unless a second FI is executed.

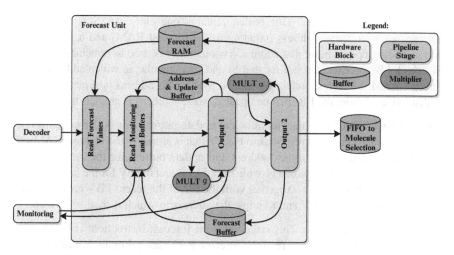

Fig. 4.12 Pipelined implementation for fine-tuning the forecasts

As discussed in Sect. 4.1, the processor pipeline needs to stall when an SI or an FI is about to execute during the fine-tuning operation of the forecast unit (see also Fig. 4.4). As an SI execution increments the monitoring value for that SI and the forecast unit resets the monitoring values for those SIs that are forecasted by an FB, inconsistencies may occur if an SI executes in parallel to a fine-tuning process of the forecasting unit. As an extreme case, resetting the FV could be prevented in the following scenario: (1) the monitoring unit reads the current monitoring value (to increment it), (2) the forecast unit resets the monitoring value, and (3) the monitoring unit writes back the incremented value, overwriting the reset activity.

The FI that triggers the execution of the forecasting unit provides the information which FB shall actually execute and how many different SIs are forecasted in this FB (i.e., how many different FVs it contains). A dedicated forecast RAM is provided that contains the data for all FBs. The FI determines a region in this forecast RAM by providing a base address and the number of entries that belong to this FB (for detailed FI instruction format, see Sect. 5.1). This separation of FI (actually a trigger) and the forecast RAM (containing the real data) provides the advantages that (1) only one instruction has to execute in the processor pipeline to trigger the fine-tuning, (2) more information can be provided (considering the number of bits) per FV than would fit into an FI, and (3) the FV can be changed without the need to update the instruction memory. In addition to the forecast RAM, two further buffers are demanded. The details for the three buffers and the four pipeline stages as they appear in Fig. 4.12 are explained.

Forecast RAM. The forecast RAM contains the actual FVs together with the information which SI is forecasted by a particular FV. In addition, it contains an "update" bit for each FV, to determine whether an FV is constant or may be updated. This RAM is initialized during compile time, depending on the FIs and FBs of the application.

Address and update buffer. This buffer contains one entry per SI.[10] Each entry corresponds to a certain address (pointing to the forecast RAM) and a copy of the corresponding update bit at that address. This memory is used to remember for each SI, which FB was executed last for that SI and, in particular, at which address in the forecast RAM the corresponding FV can be found. This address information is used to determine whether the FB shall be updated when another FB predicts the same SI.

Forecast buffer. Similar to the "address and update buffer," this buffer contains one entry per SI. This entry corresponds to the FV that is stored in the forecast RAM for the address that is stored in the "address and update buffer" for the same SI. This means that these two buffers can provide the address and the FV for the last executed FB of any SI. These values – together with the FV of the recent FB – are important to calculate the updated FV and to know the address to which it shall be written.

Read forecast values stage. This stage waits for forecast instructions (FI) from the decoder. It sets the address to read the forecast RAM according to the information of the FI. In the subsequent cycles, it increases the address according to the FB size [i.e., how many forecast values (FVs) correspond to the current forecast block (FB)] given by the FI. This stage reads the forecast RAM and provides the FVs [corresponding to $FV(FB_{t+1})$ in Eq. 4.1] for the next pipeline stage. The forecast RAM additionally provides the SI identifier for the FV. This identifier is used to access the "address and update buffer" and the forecast buffer in the next stage. When all FVs of the current FB are processed, this stage waits for the next FI.

Read monitoring and buffers stage. This stage acquires the current execution counter for the forecasted SI from the monitoring unit [corresponding to $M(FB_{t+1})$ in Eq. 4.1]. It also reads the FV of the previous forecast block FB_t from the forecast buffer [corresponding to $FV(FB_t)$ in Eq. 4.1]. In addition, it reads the address (for the forecast RAM) and update status of that previous FB from the "address and update buffer." This information is used to decide whether the FV of the previous FB shall be updated and – if yes – to which address in the forecast RAM the updated FV shall be written in the output 2 stage.

Output 1 stage. This stage resets the SI execution counter in the online monitoring and updates the "address and update buffer" with the information of the currently processed FV (information provided from "read forecast value stage"). This buffered information will be used to update the currently processed FB_{t+1} when the next FB FB_{t+2} for the same SI is reached later. This stage also calculates the error $E(FB_{t+1})$ according to Eq. 4.1. It uses the gathered information $FV(FB_t)$, $FV(FB_{t+1})$, and $M(FB_{t+1})$ from the previous stages. In the prototype, the multiplication with the parameter γ is realized as fixed-point arithmetic and accomplished by using a dedicated Virtex-4 hardware multiplier.

Output 2 stage. This stage calculates the update value for the previous forecast block FB_t, according to Eq. 4.2. A dedicated Virtex-4 hardware multiplier is used to multiply

---

[10]Altogether, 1,024 entries are reserved for all possible SI opcodes, see Sect. 5.1.

Table 4.1 Hardware requirements for monitoring and forecast unit

Slices	Flip flops	LUTs	DSP48	RAMB16
420	258	688	2	9

the parameter α. The updated FV for the previous FB is written to the forecast RAM (depending on the update bit, i.e., whether or not updating is allowed), using the address previously read from the "address and update buffer." In addition, the FV of the current FB (i.e., $FV(FB_{t+1})$) is written to the forecast buffer using the id of the SI as address. Actually, this value was read from forecast RAM and is now copied to the forecast buffer. From there it will be used to calculate the update when another FB for the same SI is executed. Providing the forecast buffer as a dedicated and redundant memory is a performance optimization. The forecast buffer could be omitted and the data could be read from the forecast RAM using the address of the previous FB provided by the "address and update buffer." However, then either the forecast RAM would need a third port (for providing the data) or the pipeline needs to be stalled to redirect one of the two ports for that operation. Eventually, this stage provides the information of the current FB to a FIFO that passes the information to the molecule selection (described in Sect. 4.4).

The described pipeline and buffers are implemented in VHDL and integrated it into the prototype. Table 4.1 provides the implementation results for the monitoring and the forecast unit. The logic requirements are moderate even though the implementation was optimized for throughput. The two DSP48 blocks correspond to the multiplications with the parameters α and γ and the RAMB16 BlockRAMs are used to realize the buffers of the forecast unit and the online monitoring.

4.3.4 Summary of the Online Monitoring and SI Forecasting

The concept of SI forecasting is used to predict which SIs are expected to be needed in the next computational block. This allows starting necessary reconfigurations in time, which is important as the reconfigurations demand relatively long (in the range of milliseconds). An online monitoring and an error back-propagation scheme are used to update the predictions according to observed SI execution requirements. This allows adapting the reconfigurations to changing SI requirements, i.e., if an SI is predicted to be executed more often (compared to previous executions of that computational block), then the novel concept of modular SIs allows choosing a faster molecule to execute it.[11] The foundation of the prediction that was described is based on the temporal difference scheme and a pipelined hardware implementation was presented and evaluated for it. The next section will describe how the expected SI execution frequencies are used as input for the so-called molecule selection to determine which molecules shall be reconfigured to implement the predicted SIs.

---

[11]Typically at the cost of slower molecules for other SIs executing in the same hot spot.

4.4 Molecule Selection

The previous section described how online monitoring and back propagation are used to predict the amount of SI executions for an upcoming computational block. This allows adapting the reconfigurable fabric toward changing SI requirements, as motivated in Fig. 3.6. The actual adaptation is performed by the so-called molecule selection [BSH08c], which is described in this section. As presented in the overview of the run-time system in Fig. 4.3 the molecule selection is triggered by the prediction, i.e., after a so-called forecast instruction (see Sect. 4.3) executed. The information provided by the prediction to the molecule selection comprises (a) which SIs are expected to execute in the next computational block, and (b) how often are the SIs expected to execute. The task of the molecule selection is to determine exactly one molecule for each requested SI.

Figure 4.13 shows a simplified selection scenario with two different SIs (SI_o, SI_p) and their corresponding molecules ($\vec{o}_i$, $\vec{p}_i$). The table on the right side of Fig. 4.13 shows the selection possibilities for different numbers of available atom containers (ACs), i.e., for different numbers of atoms that may be reconfigured at the same time. For example, if seven ACs are available, the molecules $\vec{o}_6$ and $\vec{p}_3$ may be implemented (two instances of atom A_0 and five instances of atom A_1). However, using the same number of ACs the molecules $\vec{o}_2$ and $\vec{p}_6$ may be implemented as well, which may be beneficial if the expected execution frequency of SI_p is higher than that of SI_o. Later, we analyze in more detail in which situations which molecule selection may be more beneficial than another selection. The dashed arrows in Fig. 4.13 show all four Pareto-optimal selections for both SIs, given seven ACs to implement the demanded atoms. The term "Pareto-optimal" here determines that only the largest molecules (i.e., demanding the most atoms) are shown, because

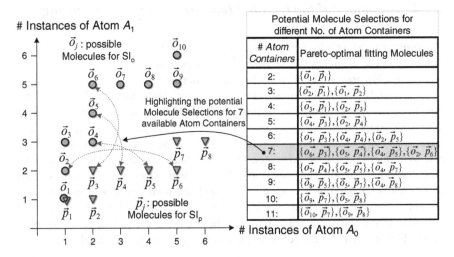

Fig. 4.13 Different showcase molecules for two special instructions with the corresponding selection possibilities for different numbers of available atom containers

the larger molecules are typically also the faster molecules (exploiting higher parallelism than smaller molecules). For instance, the molecules $\vec{o}_1$ and $\vec{p}_1$ may be selected if at least two ACs are available. However, when seven ACs are available, larger (and thus potentially faster) molecules are preferable.

The amount of available ACs might potentially change during run time, for instance, when the available reconfigurable fabric is has to be partitioned among multiple tasks (as motivated for the "SI implementation adaptivity" in Sect. 3.2). This leads to the selection alternatives shown in the table in Fig. 4.13. The problem space of determining the molecule selection grows even larger when more than two SIs are required within one computational block (which is the typical case, e.g., six SIs for the encoding engine of the H.264 benchmark, see Table 3.2). There, it is no longer beneficial to predetermine the selection at compile time. Instead, an intelligent run-time scheme is demanded to adapt the molecule selection to the changing numbers of available ACs and the SI execution frequencies efficiently.

4.4.1 Problem Description for Molecule Selection

The input to the selection is the number of available ACs N and a forecast F, which is a set of tuples $F = \{(M_i, f_i, t_i)\}$ where each tuple contains

- A forecasted special instructions SI_i [represented by a set of its molecules M_i as shown in Eq. 4.5]
- The expected execution frequency f_i f that SI (dynamically fine-tuned during run time)
- The remaining time t_i until the first execution of this SI (statically provided by offline profiling)

$$M_i = \{\vec{m}_j : \vec{m}_j.getSI() = SI_i\} \tag{4.5}$$

The task of the molecule selection is to determine a set of molecules S that has to fulfill certain conditions, described in the following. At first, it is important that the number of atoms that are demanded to implement the selected molecules do not exceed the number of available ACs, see Eq. 4.6. In addition, it is required that for each forecasted SI exactly one molecule is selected to implement it, see Eq. 4.7. As all SIs can be executed using the cISA (see Sect. 3.2) and the cISA execution does not demand any atoms [i.e., $(0, \ldots, 0)$], it is always possible to determine a valid molecule selection, even if no ACs are provided. The two conditions in Eqs. 4.6 and 4.7 describe a valid selection, but they do not consider the properties of a good selection yet. For instance, selecting the cISA execution for all forecasted SIs would always lead to a valid selection without accelerating any SI execution. The main target of a good selection is to minimize the execution time of the upcoming computational block. Each accelerating molecule may contribute to an execution-time reduction due to its speedup compared to the cISA execution. However, considering the available ACs, the molecules of different SIs need to be traded-off by evaluating their individual speedup and the expected SI execution frequency. In general, a profit is determined for each molecule and the aim is to maximize the overall profit of the selected molecules, as shown in Eq. 4.8.

$$\left| \bigcup_{\vec{o} \in S} \vec{o} \right| \le N \tag{4.6}$$

$$\forall i : |S \cap M_i| = 1 \tag{4.7}$$

$$\text{maximize} \sum_{\substack{(M_i, f_i, t_i) \in F \\ \text{with } \vec{o} := S \cap M_i}} profit(\vec{o}, f_i, t_i) \tag{4.8}$$

The details of the profit function are discussed in Sect. 4.4.2, after a complexity analysis in the remainder of this section. Considering the conditions and optimization goal described above, the selection problem has strong similarities to the knapsack problem that is known to be NP-complete [MT90]. However, some specific differences have to be examined. A knapsack problem consists of a knapsack with a given capacity C and elements $E = \{e_i\}$ with weights $w(e_i)$ and profits $p(e_i)$. The problem is to fill the knapsack with a subset of elements $E' \subseteq E$ that maximize the total profit (see Eq. 4.9) without violating the capacity of the knapsack (see Eq. 4.10). The major difference between the knapsack problem and the selection problem is the way in which the weight for a set of selected elements is determined (for the capacity constraint). In the knapsack problem, this weight is given by the sum of the weights of the selected elements (see Eq. 4.10). However, molecules may share some atoms (i.e., $|\vec{o} \cap \vec{p}| \ne 0$) and thus, the combined weight of two or more molecules is in general not equal to the sum of the weights of the individual molecules (i.e., $|\vec{o} \cup \vec{p}| \ne |\vec{o}| + |\vec{p}|$). Figure 4.14 provides an example to demonstrate this difference.

$$\text{maximize} \sum_{e_i \in E'} p(e_i) \tag{4.9}$$

$$\sum_{e_i \in E'} w(e_i) \le C \tag{4.10}$$

To proof that the selection problem is NP-hard, a polynomial reduction [GJ90] of the knapsack problem to the selection problem is given [BSH08c], as this shows that the selection problem is a generalization of the knapsack problem. Correspondingly, each instance of a knapsack problem can be solved, using a solver for the selection problem and therefore, the selection problem must be at least as hard as the knapsack problem, i.e., NP-hard. It starts with an instance of the knapsack problem as described above that is transformed into a selection problem. In the first step, the elements e_i is converted into molecules by defining one dedicated special instruction M_i per element that consists of exactly two molecules, i.e., $M_i = \{\vec{x}_i, \vec{y}_i\}$. The molecule $\vec{x}_i$ represents the cISA implementation without accelerating atoms [i.e., $(0, \ldots, 0)$] and

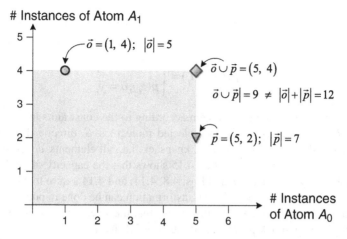

Fig. 4.14 Atom sharing, leading to a size of the combined molecule that is smaller than the accumulated size of the two individual molecules

thus, it has a zero weight. If the cISA molecule is selected, then it corresponds to the decision that the SI shall not be accelerated. For the knapsack problem this corresponds to the decision not to pack the particular element into the knapsack (see Eq. 4.11). The $\vec{y}_i$ molecule of M_i corresponds to the decision to pack the element into the knapsack. It uses a specific atom type A_i multiple times and all other atom types are not used (see Eq. 4.12). This means that an individual atom type is used for each element of the original knapsack problem. Therefore, molecules that correspond to different elements do not share any atoms (see Eq. 4.13). The quantity of instances of atom type A_i corresponds to the weight of the element e_i, i.e., $|\vec{y}_i| = w(e_i)$. Altogether, for each element of the original knapsack problem, a dedicated SI is provided that consists of two molecules that represent the decision whether or not the element is packed into the knapsack. The molecule that corresponds to the decision that the element is packed into the knapsack, uses a dedicated atom type that is not used by any other molecule and that models the weight of the element by using that atom type in the corresponding quantity.

$$\vec{x}_i \in S \Rightarrow e_i \notin E'$$
$$\vec{y}_i \in S \Rightarrow e_i \in E' \tag{4.11}$$

$$\vec{y}_i = \underbrace{(0,\ldots,0, w(e_i), 0,\ldots,0)}_{\text{atom } A_i} \tag{4.12}$$

$$i \neq j \Leftrightarrow |\vec{y}_i \cap \vec{y}_j| = 0 \tag{4.13}$$

In addition to the capacity constraint, the profit of the elements has to be considered such that they can serve as input for Eq. 4.8. Therefore, the profit of each molecule is defined as shown in Eq. 4.14 (please note that the input parameters f_i and t_i are not needed here).

$$profit(\vec{o}, f_i, t_i) = \begin{cases} 0, & \vec{o} = \vec{x}_i \\ p(e_i), \vec{o} = \vec{y}_i \end{cases} \tag{4.14}$$

After solving the selection problem according to the constraints in Eqs. 4.6 and 4.7 with $N = C$ available ACs, the selected molecules S directly determine the elements that shall be packed into the knapsack, i.e., all elements e_i for which the molecule $\vec{y}_i$ was selected. Equation 4.15 shows that the capacity of the knapsack is not violated by the selection and Eqs. 4.8, 4.11, and 4.14 assure that the profit is maximized. As the above-described transformation can be done in polynomial time (linear in number of elements e_i) this shows that the selection problem is a generalization of the knapsack problem and thus it is NP-hard.

$$\left| \bigcup_{\vec{m} \in S} \vec{m} \right| \le N = C, \text{ see Eq. 4.6} \tag{4.15}$$

$$\Leftrightarrow \quad \sum_{\vec{m} \in S} |\vec{m}| \le C, \text{ using Eq. 4.13}$$

$$\Leftrightarrow \quad \sum_{\forall i} \begin{cases} |\vec{x}_i|, \vec{x}_i \in S \\ |\vec{y}_i|, \vec{y}_i \in S \end{cases} \le C, \text{ using Eq. 4.7 and definition for } M_i$$

$$\Leftrightarrow \quad \sum_{\forall i} \begin{cases} 0, & \vec{x}_i \in S \\ w(e_i), \vec{y}_i \in S \end{cases} \le C, \text{ using Eq. 4.12 and definition for } \vec{x}_i$$

$$\Leftrightarrow \quad \sum_{\forall i} \begin{cases} 0, & e_i \notin E' \\ w(e_i), e_i \in E' \end{cases} \le C, \text{ using Eq. 4.11}$$

$$\Leftrightarrow \quad \sum_{\forall e_i \in E'} w(e_i) \le C$$

4.4.2 Parameter Identification for the Profit Function

The profit function needs to be chosen very carefully as it finally determines the quality of the resulting selection. Optimally solving the selection problem guarantees the best achievable profit but it does not guarantee the best achievable application execution time due to external constraints. For instance, the actual execution time depends on the exact SI execution frequency and the SI execution sequence. The SI execution frequency is provided as estimated parameter to the molecule selection, but as it is only a prediction, it comes with a certain error (see Sect. 4.3), which affects the relation between optimally solving the molecule selection and achieving best performance. In addition, the performance also depends on further algorithms, e.g., the reconfiguration sequence of the atoms (see Sect. 4.5) and the atom replacement (see Sect. 4.6). Nevertheless, as the molecule selection has a significant impact on the overall performance, the profit function should consider typical parameters and scenarios that may affect the application execution time. These parameters are discussed for a given molecule $\vec{m} \in M_i$ that implements SI_i. The first parameter that should be considered by the profit function is the so-called latency improvement of $\vec{m}$, i.e., the number of cycles that $\vec{m}$ is faster compared to the execution of the same SI using the cISA execution, see Eq. 4.16.

$$LatencyImprovement(\vec{m}) := \begin{array}{c} \vec{m}.getSI().getCISAMo-\\ lecule().getLatency()\\ -\vec{m}.getLatency() \end{array} \qquad (4.16)$$

$$LatencyImprovement(\vec{m}, \vec{a}) := \begin{array}{c} \vec{m}.getSI().getFastest-\\ AvailableMolecule(\vec{a}).\\ getLatency()\\ -\vec{m}.getLatency() \end{array} \qquad (4.17)$$

$$speedup(\vec{m}) := \frac{\vec{m}.getSI().getCISA-Molecule().getLatency()}{\vec{m}.getLatency()} \qquad (4.18)$$

As an alternative parameter one might consider the latency improvement in comparison to the fastest molecule that can be implemented with the currently available[12] atoms $\vec{a}$ (see Eq. 4.17). The advantage is that this parameter would consider the available atoms, i.e., a molecule would have a high profit only if it is faster than available molecules for the same SI. Therefore, if a molecule candidate is only slightly faster than the currently fastest molecule, its profit would be small, potentially giving priority to other molecules of other SIs. However, this parameter would make it complicated to explicitly select a molecule that is smaller (i.e., using less atoms) than the fastest currently existing molecule for the same SI, because all these molecules would have a negative profit value (as they are slower than the available molecule). This can cause suboptimal selections, as less atom containers are left to implement other (potentially more beneficial) SIs. Another alternative would be to use the speedup see Eq. 4.18) for the profit computation, but that may lead to disadvantages when comparing the profit values of different SIs. The cISA implementations for these different SIs may differ significantly, which is important as the comparison of the speedup is relative to the cISA execution. For instance, consider the two molecules ($\vec{o}$ and $\vec{p}$) of two different SIs shown in Example 4.3.

---

[12]That is, those atoms that are currently loaded into an atom container.

Example 4.3 Avoiding forecast thrashing by weighting the error back propagation.

- SI_1:

$$\vec{o} \in M_1; \vec{o}.getLatency() = 10;$$
$$\vec{o}.getSI().getCISAMolecule().getLatency() = 50;$$
$$speedup(\vec{o}) = 50/10 = 5\times;$$
$$LatencyImprovement(\vec{o}) = 50 - 10 = 40cycles.$$

- SI_2:

$$\vec{p} \in M_2; \vec{p}.getLatency() = 100;$$
$$\vec{p}.getSI().getCISAMolecule().getLatency() = 500;$$
$$speedup(\vec{p}) = 500/100 = 5\times;$$
$$LatencyImprovement(\vec{p}) = 500 - 100 = 400cycles.$$

As shown, two molecules might have the same speedup (5× in the example) even though they differ significantly in their respective cycle savings per execution (40 cycles vs. 400 cycles per SI execution in the example). Here it becomes noticeable that the latency improvement has a direct impact on the SI execution performance and thus the application performance. Therefore, Eq. 4.16 is considered as relevant parameter for the profit function.

Another important parameter concerns the size of a molecule and the related reconfiguration time. Bigger molecules (i.e., those with more atoms) typically exploit more parallelism and therefore may achieve a faster SI execution. This comes at the cost of longer reconfiguration time, which may reduce the overall performance of the application. To consider this scenario, the so-called reconfiguration delay is computed as the remaining cycles of reconfiguration at the time when the first SI executes, i.e., how many cycles the reconfiguration finishes too late to be useful for the first SI execution, as it is illustrated in Fig. 4.15. The starting point (i.e., all times are relative to this point) is the forecast instruction from which the molecule selection is triggered. In the case of a relatively large molecule where multiple atoms need to be reconfigured, the first SI executions may be missed, as shown in the figure. In such a case, reducing the overall profit of the molecule should be considered, because it will not be available for all SI executions and thus it will be less beneficial. In case the reconfiguration is finished in time, the profit should be neither decreased nor increase. By bounding the reconfiguration delay to be greater or equal to zero, this property is assured, see Eq. 4.19.

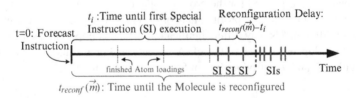

Fig. 4.15 Comparing the reconfiguration time and the first execution time of a special instruction

$$\max \left\{0, t_{\text{reconf}}\left(\vec{m}\right) - t_i \right\}; \; \vec{m} \in M_i \tag{4.19}$$

$$t_{\text{reconf}}\left(\vec{m}\right) := \left|\vec{a} \triangleright \vec{m}\right| * t_{\text{AtomReconf}}; \; \vec{a} = \text{currently avaliable atoms} \tag{4.20}$$

As already described in Sect. 4.4.1, the parameter t_i is obtained from offline profiling. Unlike the expected SI execution frequency f_i, the parameter t_i is not fine-tuned during run time. The parameter $t_{\text{reconf}}\left(\vec{m}\right)$ determines the minimal[13] possible reconfiguration time, i.e., the number of additionally required atoms (based on the currently available atoms $\vec{a}$) multiplied by the reconfiguration time of an atom, as shown in Eq. 4.20.

In addition to the two described parameters latency improvement and reconfiguration delay, the expected execution frequency of the SIs needs to be considered, which is independent from the particular molecules of a the SI. For instance, the molecules of an SI that is executed significantly more often than another SI should have a higher profit value, because a better implementation[14] of that SI has a larger impact on the overall application. The resulting profit of a molecule $\vec{m}$ for SI$_i$ is given in Eq. 4.21, where the input parameters f_i and t_i denote the expected SI execution frequency and the expected time until the first SI execution, respectively. The latency improvement and the reconfiguration delay are scaled by parameters L and R, respectively. After describing the implementation for the molecule selection, it is evaluated with respect to these parameters in Sect. 4.4.3.

$$profit\left(\vec{m}, f_i, t_i\right) := f_i \cdot \left(\begin{array}{c} L \cdot \left(\begin{array}{c} \vec{m}.getSI().getCISAMolecule(). \\ getLatency() - \vec{m}.getLatency() \end{array} \right) \\ -R \cdot \max \left\{0, t_{\text{reconf}}\left(\vec{m}\right) - t_i \right\} \end{array} \right) \tag{4.21}$$

4.4.3 Heuristic Solution for the Molecule Selection

Algorithm 4.1 Pseudo code of a greedy knapsack solver

---

1. // Input: Capacity constraint C and elements $E = \{e_i\}$ with weights $w(e_i)$ and profits $p(e_i)$, see Sect. 4.4.1.
2. // Output: subset of elements $E' \subseteq E$ that maximize the profit (see Eq. 4.9) without violating the capacity of the knapsack (see Eq. 4.10)

(continued)

---

[13]The actual reconfiguration time might be longer if the atoms of another SI are loaded in between (see Sect. 4.5).

[14]Considering latency improvement and reconfiguration delay.

Algorithm 4.1 (continued)

```
  3. ∀eᵢ ∈ E{
  4.    b(eᵢ) ← p(eᵢ)/w(eᵢ);  // calculate the so-called benefit as "profit per
                                 weight"
  5. }
  6. sort(E, b());  // sort the elements according their benefit (decreasing)
  7. E' ← ∅;  // initializes the result
  8. totalWeight ← 0;
  9. ∀eᵢ ∈ E{  // iterate according the benefit sorting
 10.    if(totalWeight + w(eᵢ) ≤ C){
 11.       E' ← E' ∪ {eᵢ};
 12.       totalWeight ← totalWeight + w(eᵢ);
 13.    }
 14. }
 15. return E';
```

Due to the complexity of the NP-hard molecule selection, a heuristic is used to solve it. Greedy implementations are a common choice for the class of knapsack problems [Hro01]. The corresponding pseudo code for the greedy approach is shown in Algorithm 4.1. The computational complexity of the greedy knapsack solver is dominated by sorting the elements (see line 6), i.e., it is $\mathcal{O}(n \log n)$ for $n = |E|$. The space requirement is determined by calculating and storing the benefit values (see line 4) that are used for sorting the elements, i.e., it is $\mathcal{O}(n)$.

The greedy implementation for the molecule selection is different from the greedy implementation for the knapsack solver, which selects the elements in descending order from a sorted list based on initially calculated "benefit" values (i.e., profit per weight). Instead of initially sorting all elements, the implementation determines the molecule with the best profit by linearly examining all candidates and then selects it. After this first selection, the remaining molecule candidates are *cleaned*, i.e., some of them might no longer fit to the available ACs (considering the already selected molecules) and some other molecules might already be implicitly available (i.e., all atoms that are needed to implement them are already available due to the atom requirements of the explicitly selected molecules). This algorithmic alternative provides benefits for this scenario even though it increases the computational complexity as discussed after presenting the algorithm. Another difference to the knapsack solver is that *exactly* one molecule has to be selected per SI (see Eq. 4.7). By implicitly selecting the cISA molecule for each SI where no other molecule was selected, that condition can be relaxed toward selecting *at most* one molecule per SI. After a molecule was selected, this property is assured by explicitly removing all molecules of that SI from the candidate list.

Algorithm 4.2 shows the pseudo code of the greedy realization of the molecule selection. After some initializations (e.g., removing the cISA molecules in lines 4–6) it enters the main loop (line 9). The algorithm computes the profits for all molecules that may be selected (i.e., that are not too big and not yet available, lines 10–23). Removing the molecules that are already available due to the atoms of the so-far selected molecules (see lines 13–16) is an optimization that reduces the number of profit calculations. To assure that these implicitly available molecules (i.e., not explicitly selected) can be used, for each SI the fastest molecule that can be implemented with the selected atoms[15] is determined after the selection completed. This will also select the cISA molecule if no faster molecule is available for a particular SI. After the profit values are calculated for each remaining molecule, the algorithm selects the best candidate (lines 25–26), and cleans the remaining candidates (lines 27–32).

Algorithm 4.2 Pseudo code of the molecule selection

1. // **Input:** number of available atom containers N and the set of forecasted SIs with the expected execution frequency and the time until the first execution: $F = \{(M_i, f_i, t_i)\}$; see Sect. 4.4.1.
2. // **Output:** set of selected molecules S, i.e., at most one molecule per forecasted SI (if no molecule is explicitly selected, then the fastest molecule that can be implemented with the atoms of the explicitly selected molecules (may be the cISA molecule) will be used
3. $M \leftarrow \emptyset$; // initializes potential molecule candidates
4. $\forall (M_i, f_i, t_i) \in F\{$ // remove cISA molecules (is chosen implicitly if no faster molecule is selected)
5. $\forall \vec{m} \in M_i, |\vec{m}| > 0 : M \leftarrow M \cup \{\vec{m}\}$;
6. $\}$
7. $\vec{a} \leftarrow (0, \ldots, 0)$;// $\vec{a}$ represents the atoms of the selected molecules
8. $S \leftarrow \emptyset$;
9. while $(\text{true})\{$ // exit condition is checked in line 24
10. $bestProfit \leftarrow minInt$;
11. $\vec{p} \leftarrow (0, \ldots, 0)$;
12. $\forall \vec{m} \in M\{$
13. // Remove the molecules that are too big or implicitly available by the so far selected atoms
14. if $(|\vec{a} \cup \vec{m}| > N \text{ or } \vec{m} \leq \vec{a})\{$
15. $M \leftarrow M \setminus \{\vec{m}\}$;
16. $\}$else$\{$
17. $prof = profit(\vec{m}, f_i, t_i)$; // see Eq. 4.21
18. if $(prof > bestProfit)\{$
19. $bestProfit \leftarrow prof$;
20. $\vec{p} \leftarrow \vec{m}$;
21. $\}$

(continued)

---

[15]That is, the atoms of the explicitly selected molecules.

Algorithm 4.2 (continued)

```
  22. }
  23. }
  24. if ($\vec{p} = (0,\ldots,0)$)break; // no molecule left
  25. $S \leftarrow S \cup \{\vec{p}\}$;
  26. $\vec{a} \leftarrow \vec{a} \cup \vec{p}$;
  27. // Remove all molecules from the same SI as the currently selected
      molecule
  28. $\forall \vec{m} \in M\{$
  29. if $(\vec{m}.getSI() = \vec{p}.getSI())\{$
  30. $M \leftarrow M \setminus \{\vec{m}\}$;
  31. }
  32. }
  33. }
  34. return$S$;
```

The required approach to recompute the profit values in the main loop (to consider the atoms of the already selected molecules) increases the complexity from $\mathcal{O}(n\log n)$ to $\mathcal{O}(n^2)$ where n is the number of molecule candidates. However, the advantage of the implementation is a reduced memory footprint, which is important when considering a hardware implementation. The memory requirements are now independent from the number of forecasted SIs and their implementing molecules as only the (profit-wise) best molecule with its profit value has to be memorized (lines 19 and 20), i.e., the memory requirements reduces from $\mathcal{O}(n)$ to $\mathcal{O}(1)$. In addition, the first selected molecule is determined after $\mathcal{O}(n)$ calculations of the profit function and thus can start the time-consuming reconfiguration earlier (in comparison to $\mathcal{O}(n\log n)$ profit function calculations for the greedy knapsack solver). In parallel to this reconfiguration, the molecule selection processes further decisions, i.e., the time for the calculations is hidden by the reconfiguration time. Altogether, the conceptual drawback of the worsened complexity turns into a performance advantage of the practical implementation.

4.4.4 Evaluation and Results for the Molecule Selection

A detailed analysis of the two scaling factors for the latency improvement L and the reconfiguration delay R is presented (see Sect. 4.4.2 and Eq. 4.21), using the H.264 video encoder benchmark presented in Sect. 3.3. To determine relevant value ranges for the scaling factors the application and the utilized SIs and molecules were analyzed. The smallest latency improvement is 20 cycles (average 148.25) and the corresponding atom loading time (indicating the potential reconfiguration delay) is 94,953 cycles[16] [BSKH07], which corresponds to nearly 1 ms at 100 MHz pipeline frequency. As these values are four orders of magnitude different from each other, they have to be normalized to make them comparable in the profit function. Without this normalization, the impact of the reconfiguration delay would dominate the latency improvement because its value is significantly larger. The latency improvement and the atom loading time are normalized to 100 cycles, respectively. For the latency improvement this corresponds to a multiplication by 5 and values for the latency improvement scaling factor L are examined accordingly (i.e., 1, 2, ..., 10). For the reconfiguration delay the normalization corresponds to a multiplication with 0.001 and values for R are examined accordingly (i.e., 0.0001, 0.0002, ..., 0.0025). Figure 4.16 shows the application execution time as a surface plot using a representative subset of the evaluated value ranges. Figure 4.16a shows the results for the greedy molecule selection and Fig. 4.16b for the optimal molecule selection, using a reconfigurable fabric with space for four ACs. The optimal molecule selection is obtained by exhaustively testing all selection combinations and choosing the one that maximizes the profit.

It is noticeable that the greedy selection nearly always reaches the performance of the optimal selection. The major difference is found in the region R1, shown in Fig. 4.16a, b. While this region is achieving the best performance for the optimal selection, the greedy selection is not able to cover this case (same for region R2). However, the greedy selection is achieving the same performance as the optimal selection in regions R3 and R4. Altogether, the greedy implementation is just lacking 11.9% performance compared to the best performance of the optimal implementation (comparing R3 and R4 of the greedy selection with R1 and R2 of the optimal selection). It is noticeable that in R5 the greedy implementation is providing a better performance than the optimal implementation. This appears to be surprising, as a heuristic should not outperform an optimal solution. However, it actually demonstrates that solving the selection problem optimally does not automatically lead to the best performance. For instance, the input for the selection problem comprises predictions about the upcoming SI execution frequencies. If this prediction turns out to be wrong[17] then an optimal selection may be outperformed by a nonoptimal solution. In addition, after solving the selection problem, further algorithms execute to eventually determine the actual atom reconfigurations (e.g., reconfiguration-sequence scheduling and atom replacement, see Sects. 4.5 and 4.6). A similar argumentation holds for the parameters of the profit function. The quality of the result depends on the parameters that where considered to determine the profit of a molecule. The molecule selection only aims to maximize that profit. However, the greedy selection only outperforms the optimal selection in a small region of parameter settings that is not relevant (because it leads to the slowest application execution in comparison to other parameter settings). Altogether, it can be observed that

---

[16]The smallest reconfiguration delay is not used here, because it is 0 if the reconfiguration finishes in time.

[17]This is the typical case; however, the error of the prediction is typically small.

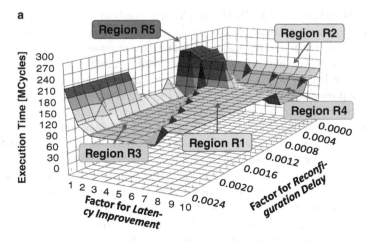

Greedy Molecule Selection for 4 Atom Containers

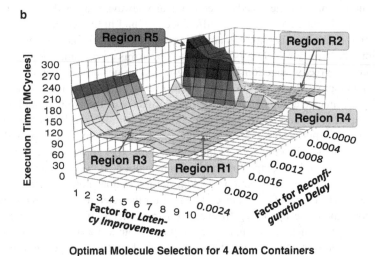

Optimal Molecule Selection for 4 Atom Containers

Fig. 4.16 Impact of the profit-function parameters on the application execution time for four atom containers

different settings for the two parameters lead to a relatively smooth surface, i.e., small parameter changes typically do not lead to significant result changes, especially the marked regions R1–R5 show that behavior.[18]

As shown in Fig. 3.5, the H.264 benchmark application comprises three different computational blocks, i.e., three different molecule selections need to be performed per frame. Therefore, the computation time of the individual blocks is analyzed in

---

[18]The shape of R1–R4 continues when examining larger values of the parameters.

more detail. Figure 4.17 shows these results as comparison between greedy selection (left side) and optimal selection (right side). It becomes apparent that the main difference between greedy and optimal selection is based on the selection results for the motion estimation (see Fig. 4.17a), which is also the computationally dominating part of the video encoder. The encoding engine unveils only two small differences in region R1 and R2. The in-loop de-blocking filter actually leads to the above-described effect that the optimal selection may result in a worse performance than the greedy selection.

After analyzing the detailed impact of the profit-function parameters on the execution time of the different computational blocks for 4 atom containers (ACs),

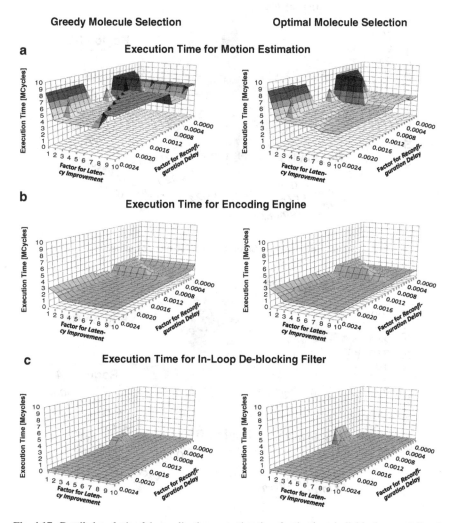

Fig. 4.17 Detailed analysis of the application execution time for the three individual computational blocks motion estimation, encoding engine, and in-loop de-blocking filter for four atom containers

the question arises how these results are affected by changing number of ACs. Potentially, the surface (i.e., values of application execution time) for different parameters might loose the property of being smooth or different parameter settings are more preferable for different number of ACs. Figure 4.18 shows the results for seven ACs, comparing greedy and optimal molecule selection. The surface is still smooth; however, region R1 is no longer significantly different between greedy and

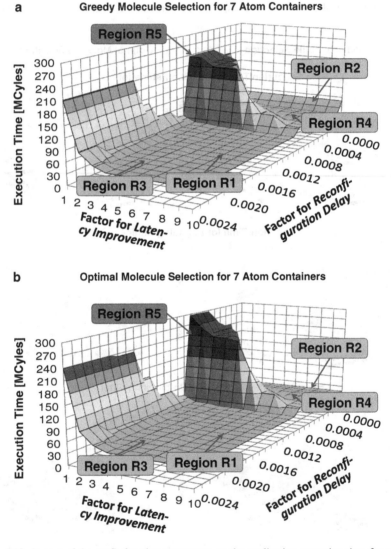

Fig. 4.18 Impact of the profit-function parameters on the application execution time for seven atom containers

optimal selection (as it was for four ACs, see Fig. 4.16). Actually, the greedy selection finds exactly the same solution that the optimal selection for the motion estimation. However, the differences that became visible for encoding engine and in-loop deblocking filter (comparing greedy and optimal) remain. The same pattern of the surface plot repeats for other numbers of available ACs. Therefore, region R3 is used for benchmarking as it provides the best solution for smaller number of ACs (see Fig. 4.16) and larger number of ACs (see Fig. 4.18). For larger number of ACs, region R1 actually leads to the fastest execution time for greedy; however, it performs badly for a small number of ACs (R3 is 1.46× faster than R1 for five ACs). For the example of seven ACs, region R1 is 1.08× faster than region R3 (only 1.05× for eight ACs).

To summarize the surface plots for other AC numbers, Fig. 4.19 gives a statistical analysis by means of box plots. Each surface plot is characterized by five values that are drawn as continuous lines for different numbers of ACs in Fig. 4.19. Each line corresponds to a so-called "x-percent" quartile. The quartiles denote certain values from the surface such that x percent of the surface values are smaller or equal to this value. In Fig. 4.19, it is noticeable that the (shaded) region between the 25- and 75% quartile is (a) very small and (b) very close to the minimum value of the surface. Both properties indicate certain robustness against changing parameter settings. Just the maximum values of the surface are outliers (i.e., significantly bigger than 75% of the values) that have to be avoided by setting the parameters accordingly (i.e., to avoid region R5). The performance for more than 5 ACs shows only small improvements (due to Amdahl's law, i.e., the performance improvement is limited by the available parallelism and the available data memory bandwidth), but still the (shaded) region between the 25- and 75% quartile gets tighter.

Figure 4.20 shows the box plots for the optimal selection. It becomes apparent that the fastest execution (0% quartile) is not significantly different compared to the greedy selection, but the slowest execution (100% quartile) is noticeably different. For small number of ACs, the slowest execution is significantly worse

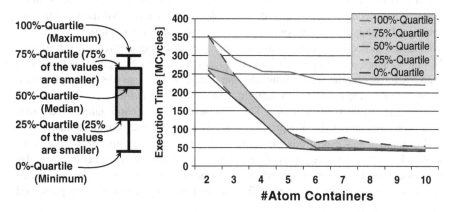

Fig. 4.19 Statistical analysis of greedy selection for different numbers of atom containers

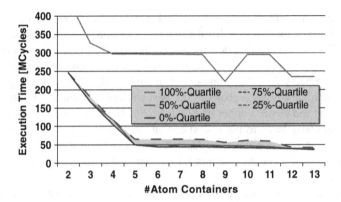

Fig. 4.20 Statistical analysis of optimal selection for different numbers of atom containers

(in comparison to the slowest greedy selection). However, in some cases, the slowest execution is faster in comparison to the greedy selection (e.g., for 9 ACs and 12 onwards). This is based on a different selection for the in-loop de-blocking filter as indicated in Fig. 4.17c. Another noticeable difference is that the shaded region between the 25- and 75% quartile is even tighter in comparison to the greedy selection. This indicates that the optimal selection is more robust against "bad" parameter settings.

4.4.5 Summary of the Molecule Selection

The molecule selection is triggered for each computational block by forecast instructions. It determines molecules to implement the forecasted SIs such that all these molecules fit to the reconfigurable hardware and a profit function – considering the SI execution frequency and the latency improvement and reconfiguration delay of a molecule – is maximized. The molecule selection is NP-hard and thus an optimized greedy approach was developed to solve it. This is particularly important, as the molecule selection has to be calculated during application run time to be able to adapt on changing SI execution frequencies and changing availability of reconfigurable fabric. Evaluating the parameter settings of the profit function showed that it is rather stable to changing scenarios and provides a good application performance in most of the cases. The comparison with an optimally solved molecule selection shows that the proposed greedy approach performs reasonably well and in many cases even finds the same solution than the optimal selection. After molecules are selected for the forecasted SIs, the atoms that are needed to implement them need to be reconfigured. As only one reconfiguration may be performed at a time, the sequence of the reconfigurations has to be determined, as presented in the next section.

4.5 Reconfiguration-Sequence Scheduling

After the molecule selection (see Sect. 4.4) has determined which molecules shall be reconfigured to implement the forecasted SIs (see Sect. 4.3), the reconfigurations of the demanded atoms need to be started. A constraint of all existing FPGA platforms is that at most one reconfiguration can be performed at a time and therefore, the demanded atom reconfigurations need to be scheduled, i.e., a reconfiguration sequence needs to be determined [BSKH08]. Future FPGA families may potentially allow performing two or more reconfigurations in parallel; however, this comes with significant conceptual drawbacks. The reconfiguration time depends on the amount of configuration bits that have to be written (the so-called bitstream) and the memory bandwidth to access the reconfiguration data memory, i.e., $t_{reconf}[ms] = Bitstream[KB] / ReconfBandwidth[KB/ms]$ (note: $x[KB/ms]$ directly correspond to $x[MB/s]$). When using a given reconfiguration bandwidth to perform two reconfigurations, then it is not beneficial to perform them in parallel (assuming that two reconfiguration ports would be available) as the available bandwidth needs to be shared between both reconfigurations. Partitioning the available bandwidth between multiple reconfiguration ports does not affect the duration until all reconfigurations are completed. However, using the full available bandwidth to perform one reconfiguration after the other leads to the advantage that the first reconfiguration is completed as early as possible (considering the given bandwidth) and therefore some atoms may be used rather early. Then the question arises, which atom shall be reconfigured first etc., i.e., the reconfiguration sequence.

The importance of a good reconfiguration-sequence scheduling is illustrated by a simple example in Fig. 4.21. The figure shows three different molecules $\vec{m}_i$ that implement the same SI. The molecule selection determined that the molecule $\vec{m}_3$ shall be used to implement the SI. A good schedule should exploit the potential for

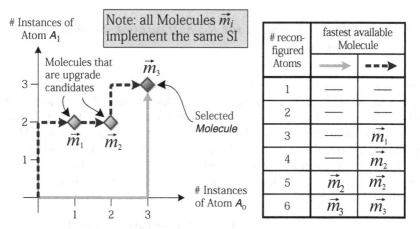

Fig. 4.21 Different atom schedules with the corresponding molecule availabilities

upgrading from one molecule to a faster one until the selected molecule is finally composed. Without this upgrading, the SI cannot utilize the accelerating atoms for a noticeable longer time and has to use the rather slow cISA execution instead (see Sect. 5.2). Figure 4.21 shows two different schedules that perform six reconfigurations to eventually implement the selected molecule $\vec{m}_3$. The schedule that is indicated by the green continuous line first loads all instances of atom A_0 and afterwards all instances of atom A_1. After four reconfigurations the point $(3,1)$ is reached (i.e., three instances of atom A_0 and one instance of atom A_1), which is not sufficient to implement any of the available molecules. After the fifth reconfiguration, $\vec{m}_2$ becomes available and thus the SI implementation is upgraded from the cISA execution to a hardware execution using the atoms. Please note that also $\vec{m}_1$ becomes available after the fifth reconfiguration; however, as it exploits less parallelism (using less atoms) it is slower than $\vec{m}_2$ and thus $\vec{m}_2$ is preferred. The schedule that is indicated by the dark dashed line exploits the potential to SI upgrades. At first, the atoms that are needed to implement $\vec{m}_1$ are reconfigured and thus already after three reconfigurations the SI implementation is upgraded and the slow cISA execution is not demanded anymore. Afterwards, the schedule upgrades to $\vec{m}_2$ and eventually to $\vec{m}_3$.

This schedule cannot be determined at compile time because it highly depends on run-time properties. For instance, the starting point of the schedule (which atoms are currently available) and the ending point (which molecule is selected depending on the amount of available atom containers and the predicted SI execution frequency) depend on the run-time situation. Typically, multiple SIs are predicted for one computational block (see, for instance, Fig. 3.5) and thus the scheduler has to decide which SI to upgrade first. In addition, these different SIs may demand multiple different atom types (not only two as shown in the example in Fig. 4.21) and molecules of different SIs may share some atoms, which enlarges the problem space of determining the atom-loading sequence. A too simplistic reconfiguration-sequence scheduling will not exploit the full performance potential of the RISPP system, as shown later.

4.5.1 Problem Description for Reconfiguration-Sequence Scheduling

The input to the reconfiguration-sequence scheduling is a set $M = \{\vec{m}_i\}$ of all molecules $\vec{m}_i \in \mathbb{N}^n$ that are selected for implementation. The meta-molecule $\sup(M)$ (see Eq. 3.7 and Fig. 3.11) contains all atoms that are needed to implement the selected molecules. The number of instances of the ith atom of $\sup(M)$ is named as "x_i" (see Eq. 4.22). NA is defined as the number of atoms that are needed to implement all selected molecules (see Eq. 4.23). The previously executed molecule selection guarantees that these atoms fit to the available atom containers (ACs), i.e., $\mathrm{NA} \leq \#\mathrm{ACs}$.

$$\sup(M) = (x_1, x_2, \ldots, x_n); x_i := \frac{\#\text{Demanded instances of atom } A_i}{\text{to implement the selected molecules}} \quad (4.22)$$

$$\text{NA} := \left|\sup(M)\right| = \sum_{i=0}^{n-1} x_i \tag{4.23}$$

With these definitions, a loading sequence for atoms can be defined as a scheduling function SF as shown in Eq. 4.24. The interval $[1,k]$ hereby represents k consecutive moments where in the moment $j \in [1,k]$ the reconfiguration of the atom $A_i = \text{SF}(j)$ is started. The variable k determines the amount of demanded reconfigurations and thus depends on the currently available atoms $\vec{a}$ and the selected molecules M (see Eq. 4.25). The scheduling function for the schedule that is indicated by the dark dashed line in Fig. 4.21 is $\text{SF}(1) = A_1$, $\text{SF}(2) = A_1$, $\text{SF}(3) = A_0$, $\text{SF}(4) = A_0$, $\text{SF}(5) = A_1$, $\text{SF}(6) = A_0$.

$$\text{SF} : [1,k] \rightarrow \{A_0, \ldots, A_{n-1}\} \tag{4.24}$$

$$k := \left|\vec{a} \triangleright \sup(M)\right| \tag{4.25}$$

$$\forall i \in [0, n-1] : \left|\{j \mid \text{SF}(j) = A_i\}\right| = x_i \tag{4.26}$$

$$\sum_{j=1}^{k} T(\text{SF}(j)) = \sup(M) \tag{4.27}$$

To make sure that exactly those atoms are loaded which are needed to implement the requested molecules (i.e., $\sup(M)$) an additional condition needs to be defined that restricts the general scheduling function of Eq. 4.24. The condition in Eq. 4.26 ascertains, that SF considers each atom A_i in the correct multiplicity x_i. The variable i makes sure that all atom types A_i are considered and the variable j examines all k reconfigurations whether atom A_i is reconfigured during that moment. All these moments are collected in a set and the determinant of that set has to equal the demand number of instances for atom A_i. Altogether, this guarantees the property shown in Eq. 4.27. The scheduling function in Eq. 4.24 with the condition in Eq. 4.26 describes a *valid* schedule. Now the properties of a *good* schedule have to be discussed. The main goal is to reduce the number of cycles that are required to execute the upcoming computational block (for which the molecule selection determined the molecules that shall be reconfigured). Therefore, it is essential to exploit the architectural feature of stepwise upgrading from slower to faster molecules until all selected molecules are available. An *optimal* schedule would require a precise future knowledge (e.g., which SI will be executed when) to determine the schedule that leads to the fastest execution. For a realistic scheduler implementation, less exhaustive future knowledge needs to be assumed. Based on the online monitoring (see Sect. 4.3) an estimate, which SI is more *important* (in terms of expected executions) than another one is available, i.e., the expected SI execution frequency is obtained as additional input.

As shown in the example in Fig. 4.21, major performance changes occur then, when an SI can upgrade from an available molecule to a faster one. Therefore, the problem of scheduling atoms is reduced to the problem of scheduling molecules. This strategy not only reduces the scheduling complexity (each scheduled molecule requires at least one additional atom) but it also allows for a clear expression which SI shall be upgraded next. Out of a molecule reconfiguration sequence, the atom reconfiguration sequence has to be determined. When $\vec{a}$ denotes the available (or already scheduled) atoms and $\vec{m}$ denotes the molecule that shall be scheduled next, the atoms in the meta-molecule $\vec{a} \triangleright \vec{m}$ need to be reconfigured in the next $|\vec{a} \triangleright \vec{m}|$ steps. When $i \in [0, k-1]$ atom reconfigurations were already determined, then scheduling function need to fulfill the condition from Eq. 4.28. The sequence in which the atoms from the meta-molecule $\vec{a} \triangleright \vec{m}$ shall be reconfigured is not explicitly expressed here, because this sequence is not important to upgrade the targeted SI toward the envisioned molecule $\vec{m}$.

$$\sum_{j=1}^{|\vec{a} \triangleright \vec{m}|} T\left(SF(i+j)\right) \overset{!}{=} \vec{a} \triangleright \vec{m}, i \in [0, k-1] \tag{4.28}$$

Not all molecules $\vec{m}$ are allowed to be scheduled in that way, only the molecules that are upgrade candidates to the molecules that are determined by the molecule selection shall be considered here. If a molecule that is not such an upgrade candidate would be scheduled, then potentially Eq. 4.27 [and thus the actual condition from Eq. 4.26] would be violated. Instead, the focus has to be placed on those molecules that are upgrade candidates for the selected molecules of the predicted SI. To make use of stepwise upgrading from slower to faster molecules, at first all smaller molecules M' that implement the same SIs as the selected molecules M need to be determined (see Eq. 4.29[19]).

$$M' = \bigcup_{\vec{m} \in M} \left\{\vec{o} : \vec{o} \leq \vec{m} \wedge \vec{o}.getSI() = \vec{m}.getSI()\right\} \tag{4.29}$$

The molecules in M' are all possible intermediate steps that might be considered on a schedule that eventually reaches $\sup(M) = \sup(M')$. However, some further optimizations have to be considered. A schedule candidate (i.e., a molecule $\vec{m} \in M'$) might be already available (i.e., $\vec{m} \leq \vec{a}$) although it has not been explicitly scheduled. This depends on the atoms in $\vec{a}$ that were (a) initially available in the atom containers or (b) atoms of those molecules that are already scheduled. Such a molecule does not need to be scheduled explicitly, because it would not trigger any atom reconfigurations, i.e., it would not change the scheduling function SF. Furthermore, a currently unavailable molecule[20] is not necessarily faster than the currently fastest available

---

[19]Please note that the union operator in (4.29) is a set union, not a molecule union, i.e. M' is a set of molecules not a meta-molecule

[20]That is, not all demanded atoms are available.

(or scheduler) molecule for the same SI. For example, Fig. 4.21 may contain a third upgrade candidate $\vec{m}_4 = (1,3)$ with a worse latency[21] than $\vec{m}_2 = (2,2)$ but a faster latency than $\vec{m}_1 = (1,2)$. After $\vec{m}_2$ is composed by the schedule with the dashed line (see Fig. 4.21), $\vec{m}_4$ is still unavailable and it does not offer a latency improvement. Therefore, such a molecule is not scheduled explicitly. However, such a molecule may be beneficial in certain scenarios, depending on the initially available atoms $\vec{a}$. In Fig. 4.21, it is assumed that initially no atom is available, but – as discussed in the context of the figure – the available atoms depend on the reconfigurations for the previously executed computational blocks. In this example, $\vec{m}_4$ may be beneficial if $\left| (\vec{a} \triangleright \vec{m}_4) \right| \leq \left| (\vec{a} \triangleright \vec{m}_2) \right|$, e.g., for $\vec{a} = (0,3)$. In this example, $\vec{m}_4$ is an upgrade candidate, because it can be reached with one reconfiguration and leads to a faster latency than $\vec{m}_1$. Therefore, it cannot be assumed that molecules like $\vec{m}_4$ are removed at compile time, as they may be beneficial depending on the currently available atoms. Instead, the list of molecule candidates from M' is cleaned by removing those molecules that are already available (as discussed above) and those atoms that do not lead to a performance improvement in comparison to the currently fastest molecule of the same SI (see Eq. 4.30).

$$M'' = \left\{ \vec{m} \in M' : \left(\begin{array}{c} |\vec{a} \triangleright \vec{m}| > 0 \wedge \vec{m}.getLatency() < \\ \vec{m}.get\ SI().getFastestAvailable\text{-} \\ Molecule(\vec{a}).getLatency() \end{array} \right) \right\} \qquad (4.30)$$

From the molecules in M'' one particular molecule is chosen (different policies are presented in Sect. 4.5.2) and the scheduling function to reconfigure the additionally demanded atoms is defined. Afterwards, the remaining scheduling candidates are cleaned according to Eq. 4.30 (using M'' as input for Eq. 4.30 instead of M' that was used as input for the initial cleaning).

4.5.2 Determining the Molecule Reconfiguration Sequence

Different strategies to determine the molecule reconfiguration sequence are presented, and their advantages and disadvantages are discussed [BSKH08]. The examples shown in Fig. 4.22 are used to explain the strategies. The figure shows two SIs (circles for SI_1 and squares for SI_2) with multiple molecules per SI. To ease the explanation, the discussion is restricted to molecules that only use two different atom types (A_0 and A_1). The two dark-filled molecules ($\vec{m}_1$ and $\vec{m}_2$) are the molecules that are selected to implement to two SIs. The lighter-filled molecules are intermediate upgrade possibilities for the SIs. It is considered, that initially no atoms are available (i.e., $\vec{a} = (0,\ldots,0)$). The target of the scheduling strategy is to load those atoms that both selected molecules are available, i.e., that $\sup(\{\vec{m}_1, \vec{m}_2\})$ is reached.

---

[21]That is, demanded number of cycles for one execution of the molecule.

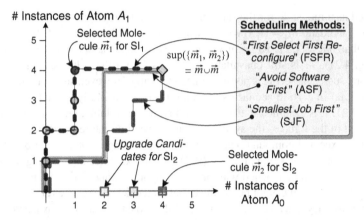

Fig. 4.22 Comparing different scheduling methods for two selected molecules of different SIs

First select first reconfigure (FSFR). As the possibility to upgrade from one molecule to a faster one can lead to a significant improved execution time (as shown in Fig. 4.21), FSFR mainly concentrates on exploiting this feature. In general, multiple SIs are required to accelerate a computational block. FSFR concentrates on first upgrading the most *important* SI until it reaches the selected molecule, before starting the second most important SI. The term "important" denotes the sequence in which the molecule selection (see Sect. 4.4) decides which molecule shall be used to implement an SI. The selection algorithm calculates a profit value for each molecule of all demanded SIs and selects the molecule with the highest profit first. This iterates until for all SIs a molecule is selected, i.e., there is a sequence in which the SI implementations are selected, and this sequence reflects the result of the profit function. By performing the SI upgrades in the same sequence, the parameters that are used to determine the profit of the molecule that is selected first are considered. However, this strategy upgrades one SI after the other (up to the selected molecule, respectively). It is not possible to upgrade one SI to some degree (not up to the selected molecule), then upgrade another SI, and eventually continue upgrading the first SI.

Avoid software first (ASF). One potential problem of the FSFR schedule in Fig. 4.22 is that SI_2 is not accelerated by a molecule until SI_1 is upgraded up to the selected molecule. Typically, the first upgrade steps that move from the cISA execution to a hardware execution lead to the highest latency improvement, whereas later upgrade steps sometimes only lead to a molecule latency reduction of a few cycles. Therefore, after a reasonable fast hardware molecule for SI_1 is available, upgrading it further does not necessarily lead to an overall application performance improvement, because SI_2 is sill executed in the relatively slow cISA implementation which now dominates the execution time. Therefore, the ASF scheduler concentrates on first loading an accelerating molecule for all SIs to avoid the relatively slow cISA execution (first phase). Again, it uses the sequence that was determined by the molecule selection to upgrade all SIs that still use the cISA execution. After no cISA execution is demanded any more, the remaining molecule updates are performed using the FSFR algorithm, i.e., one SI after the other is upgraded up to the selected molecule (second phase).

Smallest job first (SJF). Continuing the idea of loading small molecules first, leads to the SJF schedule. It starts with the same strategy like the ASF algorithm, i.e., it assures that no SI demands the cISA molecule for execution any more (first phase). Afterwards, instead of following the FSFR approach, for all remaining molecule candidates the number of additionally required atoms is determined and the molecule with the minimal additional atoms is selected, i.e., the additionally demanded atoms are scheduled (second phase). If two or more molecules require the same minimal number of additional atoms, then the molecule with the bigger performance improvement (i.e., latency reduction) is scheduled first. For instance, in Fig. 4.22, after the SJF schedule established a hardware molecule for both SIs (i.e., at position $(2,1)$), two possibilities exist to upgrade an SI implementation by just loading one further atom. If A_0 is reconfigured then SI_2 is upgraded, if A_1 is reconfigured then SI_1 is upgraded. Depending on the latency improvement (not shown in the figure), SJF decides to upgrade SI_1 first.

Highest efficiency first (HEF). All three presented schedulers bear certain drawbacks. Either they concentrate on upgrading one SI after the other (FSFR, ASF in the first and second phase, SJF in the first phase), or they concentrate on selecting the locally smallest upgrade step (SJF in the second phase). What is needed is a scheme that situation-dependent determines whether it is more beneficial to continue upgrading a certain SI or to switch to a different SI and later on continue the previous SI. For instance, consider the SJF strategy for SI_o and SI_p in the example shown in Fig. 4.23. After reconfiguring one molecule (to avoid the cISA execution) for both SIs (i.e., reaching molecule $\vec{P}_1$), SJF considers the smallest upgrade step.

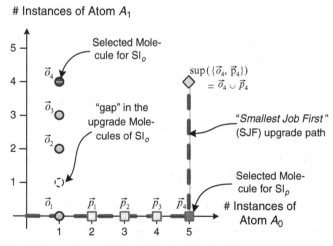

Fig. 4.23 The problem of upgrade "Gaps" for the SJF scheduler

As SI_o has a *gap* in its upgrade chain (i.e., there is one upgrade molecule that demands two or more additional atoms) SJF concentrates on SI_p first. However, SI_o might be very often executed and thus become a performance bottleneck if it is not upgraded early. Even if SI_o would not have such an upgrade gap, it still depends only on the latency improvement of the next upgrade step, which molecule SJF will choose. For instance, if SI_o offers another molecule $\vec{o}_5 = (1,1)$ but this molecule only provides minor (or even none) latency improvement in comparison to $\vec{o}_1$, then SJF would not choose it. However, this upgrade step might enable further upgrades (i.e., $\vec{o}_2$, etc.) that then only need one additional atom and that might provide major performance improvements. Therefore, a metric is needed that considers these situations and that determines the upgrade molecule that are the most beneficial ones on a scheduling path up to $\sup(M)$. This metric needs to consider the latency improvement, the SI execution frequency, and the amount of additionally demanded atoms. Please note that – unlike the latency improvement that is used for the molecule selection (see Eq. 4.16 in Sect. 4.4.2) – the latency improvement relative to the currently fastest molecule of that SI is considered (considering the already scheduled atoms) instead of the latency improvement relative to the cISA execution.

Algorithm 4.3 The implemented scheduling method "highest efficiency first"

1. // **Input:** set of selected molecules $M = \{\vec{m}\}$ and predicted SI execution frequencies $f[SI_i]$ from online monitoring.
2. // **Output:** Sorted list of molecules *scheduledList*, i.e., molecule schedule.
3. // Consider all *smaller molecules*, see Eq. 4.29
4. $M' \leftarrow \varnothing$;
5. $\forall \vec{m} \in M \{$
6. $\forall \vec{o} : \vec{o} \leq \vec{m} \wedge \vec{o}.getSI() = \vec{m}.getSI() \{$
7. $M' \leftarrow M' \cup \{\vec{o}\}$;
8. }
9. }
10. // Initialize the *bestLatency* array for all forecasted SIs
11. $\vec{a} \leftarrow$ currentlyAvailableAtoms;
12. $\forall \vec{m} \in M \{$
13. $bestLatency[\vec{m}.getSI()] \leftarrow$ $\vec{m}.getSI().getFastestAvailable\text{-}$ $Molecule(\vec{a}).getLatency()$;
14. }
15. // Schedule the molecule candidates
16. $scheduledList.clear()$;
17. $while(M' \neq \varnothing)\{$
18. // Clean molecule candidates, see Eq. 4.30
19. $\forall \vec{m} \in M' \{$
20. $if(\vec{m} \leq \vec{a} \vee \vec{m}.getLatency() \geq bestLatency[\vec{m}.getSI()])$

(continued)

Algorithm 4.3 (continued)

21. $M' \leftarrow M' \setminus \{\vec{m}\}$;
22. }
23. if $\left(M' = \varnothing\right) break;//$ no more upgrade candidates unavailable or (latency-wise) beneficial
24. $bestBenefit \leftarrow 0$;
25. $\forall \vec{o} \in M' \{$
26. $benefit \leftarrow \dfrac{\vec{o}.getSI().getExpectedExecutions()}{} * \left(bestLatency\left[\vec{o}.getSI()\right] - \vec{o}.getLatency()\right) / |\vec{a} \triangleright \vec{o}|$;
27. if $\left(benefit > bestBenefit\right)\{$
28. $bestBenefit \leftarrow benefit$;
29. $\vec{m} \leftarrow \vec{o}$;
30. }
31. }
32. // Schedule the chosen molecule
33. $scheduledList.push\left(\vec{m}\right)$;
34. $\vec{a} \leftarrow \vec{a} \cup \vec{m}$;
35. $bestLatency\left[\vec{m}.getSI()\right] \leftarrow \vec{m}.getLatency()$;
36. }
37. return $scheduledList$;

Algorithm 4.3 shows the pseudo code of the proposed HEF scheduling algorithm. At first, it collects all upgrade candidates [lines 4–9, as explained in Eq. 4.29] and initializes an internal data structure that maintains the fastest SI execution latency considering the currently available atoms (lines 11–14). Later, this data structure will be upgraded after a molecule is scheduled (line 35). In the main loop (lines 17–36) the upgrade candidates are cleaned [according to Eq. 4.30, see lines 19–22] and the molecules with the highest benefit is determined (lines 24–31). The benefit computation for the molecule candidates is shown in line 26. The performance improvement compared to the currently fastest available/scheduled molecule for the same SI is weighted with the number of expected SI executions and the number of additionally required atoms. Eventually, the molecule with the highest benefit is pushed to the *scheduledList* (the result of the algorithm) and the data structures are updated (lines 33–35).

The *scheduledList* (the result of the HEF algorithm in Algorithm 4.3) contains the molecules in the sequence as they shall be reconfigured. However, this sequence may contain gaps, i.e., HEF does not assure that all possible upgrade steps are considered. Instead, HEF determines milestones that should be targeted, not specifying how these milestones shall be reached. To fill these potential gaps between the milestones further upgrade molecules are inserted into the HEF schedule. Whenever the HEF scheduler decided that molecule $\vec{m}$ should be reconfigured next, then all upgrade candidates $\vec{o} \le \vec{m}$ of the same SI $(\vec{o}.getSI() = \vec{m}.getSI())$ that are not yet available $(|\vec{a} \rhd \vec{m}| > 0)$ are inserted into the *scheduledList*. This basically corresponds to the FSFR policy, because only one SI is upgraded. From this extended *scheduledList* the scheduling function SF is generated by scheduling those atoms that are additionally demanded to implement the next molecule of the *scheduledList* (iterating over all molecules of the *scheduledList*).

4.5.3   *Evaluation and Results for the Reconfiguration-Sequence Scheduling*

The four different proposed scheduling algorithms are benchmarked using the H.264 video encoder and SIs from Sect. 3.3 for encoding 140 frames of a CIF[22] video sequence with different atom container (AC) quantities. The results for 0–4 ACs are omitted for clarity, because their execution time is significantly longer[23] and the impact of the chosen scheduling algorithm does not become apparent if such few atom reconfigurations need to be scheduled. Instead, a noticeable situation can be seen when seven ACs are available. The performance for the FSFR (first select first reconfigure), ASF (avoid software first), and later also SJF (smallest job first) scheduler degrades when more ACs are added. This is due to the fact, that bigger molecules (i.e., molecules with more atoms) shall be reconfigured (determined by the molecule Selection from Sect. 4.4 as more space is available) and this increases the reconfiguration time until the selected molecules are finally available. Therefore, although these bigger molecules offer the potential for a faster execution, this potential has to be made available by a more sophisticated scheduling scheme. Especially FSFR fails here, as it strictly upgrades one SI after the other. However, from 17 ACs on, FSFR outperforms ASF, as ASF initially spends some time to accelerate all SIs, even though some of them are significantly less often executed than others are. The HEF scheduling does not underlie such drawbacks, as it is able to weight the importance of the molecules independently. Therefore, it always leads to the best performance in comparison to the other scheduling schemes. The more ACs are available, the clearer the differences between the scheduling methods become apparent.

Figure 4.24 focuses on showing the differences between the schedulers and therefore a zoom on the *x*-axis is provided (i.e., the *x*-axis does not start at zero). To provide a relative comparison of the different schedulers, Table 4.2 shows the speedup of the HEF scheduler in comparison to the FSFR, ASF, and SJF scheduler for different amount of atom containers. In comparison to FSFR, HEF provides up

---

[22]Common intermediate format, i.e., 352×288 pixels.

[23]Slowest execution speed for zero ACs (corresponds to a general-purpose processor): 7,403 million cycles.

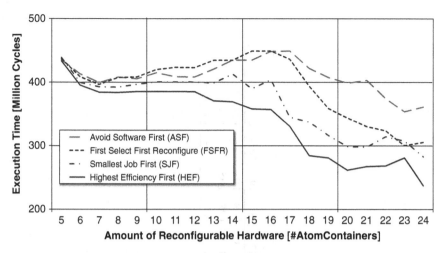

Fig. 4.24 Comparing the proposed scheduling schemes for different amount of atom containers

to 1.38× speedup (for 18 ACs), in average 1.17× for all 20 AC quantities (5–24 ACs) and in average 1.26× for 15–24 ACs. In comparison to ASF, HEF provides up to 1.52× speedup (for 20 and 24 ACs), in average 1.23× for all 20 AC quantities and in average 1.40× for 15–24 ACs. In comparison to SJF, HEF provides up to 1.19× speedup (for 18 and 24 ACs), in average 1.08× for all 20 AC quantities and in average 1.13× for 15–24 ACs. This shows that the impact of a good scheduling decision is higher when more ACs are available.

The detailed scheduling behavior of HEF is analyzed by illustrating the direct scheduling decisions and the resulting performance changes. Figure 4.25 shows the first two computational blocks of the H.264 video encoder (i.e., motion estimation and encoding engine as illustrated in Fig. 3.5) executed for one frame using ten atom containers. The x-axis shows a time axis from 0 to 2.4 million cycles after the application started. The lines show the latencies for four SIs on a logarithmic scale (see left y-axis) and thus the immediate scheduler decision. Whenever a latency line decreases, the atoms to upgrade the molecule just finished loading. The bars show the resulting SI execution frequency for periods of 100,000 cycles (see right y-axis), thus showing the performance improvement due to the scheduling.

The first scheduling decision upgrades SAD, which improves the performance significantly from 4,673 cycles (cISA execution) to 68 cycles (smallest hardware molecule). Afterwards the HEF scheduler upgrades SATD multiple times, and then performs one more upgrade for SAD and eventually two minor upgrades for SATD. At the beginning of the execution, rather few SAD and SATD SI are executed per 100,000 cycles (i.e., the bars are rather short). During the SI upgrades, more SIs get executed and after all upgrades completed, the SI execution frequencies stabilize at a relatively high value. After the motion estimation completed, the SIs for MC and DCT are upgraded. Again, it becomes visible how the HEF scheduler

Table 4.2 Speedup due to HEF scheduling

#Atom Containers	5	6	7	8	9	10	11	12	13	14	15	16	17	18	19	20	21	22	23	24
HEF vs. FSFR	1.01	1.03	1.03	1.06	1.06	1.09	1.10	1.10	1.17	1.18	1.25	1.26	1.32	1.38	1.28	1.31	1.23	1.21	1.07	1.29
HEF vs. ASF	1.00	1.04	1.04	1.06	1.05	1.08	1.06	1.06	1.13	1.18	1.21	1.26	1.36	1.48	1.45	1.52	1.51	1.39	1.26	1.52
HEF vs. SJF	1.01	1.01	1.02	1.02	1.03	1.04	1.04	1.04	1.08	1.12	1.09	1.13	1.04	1.19	1.13	1.14	1.11	1.17	1.09	1.19

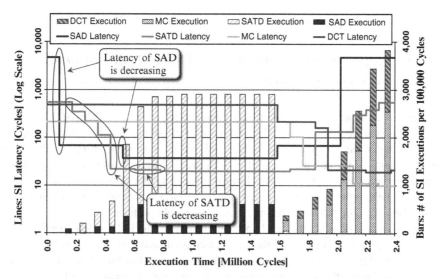

Fig. 4.25 Detailed analysis of the HEF scheduler for the motion estimation and encoding engine, showing how the SI latencies (*lines*) and execution frequencies (*bars*) change over time

switches between the SIs, depending on which upgrade step seems more benefi-cial. During the upgrades for MC and DCT, the latency for SAD and SATD gets worse, i.e., they are downgraded. This is because free atom containers are demanded to upgrade MC and DCT and thus some atoms for SAD and SATD are replaced (see Sect. 4.6).

4.5.4 Summary of the Reconfiguration-Sequence Scheduling

The reconfiguration-sequence scheduling determines the sequence in which atoms shall be reconfigured. It is triggered after the molecule selection, i.e., when new molecules shall be reconfigured for an upcoming computational block. This sequence is performance-wise important as (a) at most one atom can be reconfigured at the same time and (b) the atom reconfiguration sequence determines which mol-ecules are available and thus whether or not the feature of molecule upgrading is exploited. The task of determining the atom reconfiguration sequence is reduced to the task of determining the next molecule upgrade and presented four different strat-egies to solve this problem. The HEF scheduler performed better in all scenarios in comparison to the other algorithms. It considers the SI latency improvement, the SI execution frequency, and the amount of additionally demanded atoms to determine the next molecule upgrade. To reconfigure a new atom, a free atom container (AC) is needed for it. If no ACs are free, some atoms need to be replaced, as presented in the next section.

4.6   Atom Replacement

4.6.1   Motivation and Problem Description of State-of-the-Art Replacement Policies

Whenever an atom shall be reconfigured and no free atom container (AC) is available, then one atom has to be replaced. Free ACs might be available after the application started and did not reconfigure all ACs, yet. Depending on the AC requirements of the application, all free ACs are used (i.e., reconfigured to contain an atom) after the first forecasts. Before presenting the developed replacement policy, an example for a typical replacement situation is presented and discussed to show how state-of-the-art replacement policies would perform.

Figure 4.26 shows a high-level application flow of an H.264 video encoder (see also Sect. 3.3) that consists of three major computational blocks, namely motion estimation (ME), encoding engine (EE), and in-loop de-blocking filter (LF). They execute subsequently per video frame and use multiple computationally intensive special instructions (SIs), e.g., SAD, SATD, DCT, and HT. A time budget of 33 ms per frame (targeting 30 frames per second) allows multiple reconfigurations per frame. After the SIs of ME finished execution, the reconfigurations for the SIs of EE may start. This may demand replacements of available atoms from the reconfigurable fabric. Note, that the typical execution time of a computational block and the hardware requirements of its SIs differ for ME, EE, and LF (as indicated in Fig. 4.26).

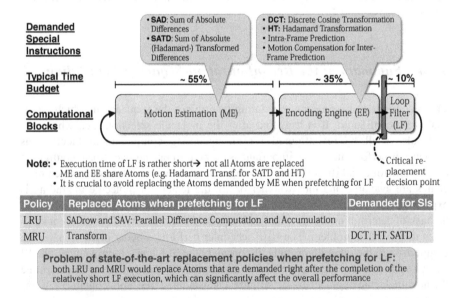

Fig. 4.26 High-level H.264 video encoder application flow, showing a typical use case and presenting the different replacement decisions of LRU and MRU in detail

For instance, LF typically has neither the time window[24] nor the hardware requirements to reconfigure large parts of the reconfigurable fabric. Therefore, some of the available[25] atoms remain in the reconfigurable fabric. When these atoms are demanded again, they can be used without additional reconfiguration. Here, the replacement policy has to decide which atoms shall remain available in the reconfigurable fabric for potential future use.

There are noticeable differences between the replacement requirements for reconfigurable processors and for cache lines and memory pages:

- Cache and page replacement policies [Tan07] aim at exploiting the locality of memory accesses. In reconfigurable processors, the locality of SI executions – and thus, the requirements of the corresponding atoms – is predetermined by the prefetching mechanism (see Sect. 4.3), independent of whether prefetching is decided upon statically or dynamically [LH02]. This means that prefetching predicts which SIs[26] and atoms[27] are most beneficial for a computational block, and which are most likely not demanded (those are replacement candidates).
- Furthermore, the Belady replacement policy [Bel66] (optimal for cache and page replacement but demanding future knowledge) is not necessarily advantageous for reconfigurable processors. This policy necessitates knowledge about the upcoming memory accesses and replaces those cache lines that are not needed for the longest time into the future. As future knowledge is typically not available, this policy is not practically implementable, but it determines the optimal replacement policy for cache and page replacements. However, for reconfigurable processors the situation is different. An access to a memory page that is currently not loaded has to be stalled until that page is swapped in. Instead, atoms are not necessarily *required* for application execution. For instance, a reconfigurable processor that executes an SI can still perform by using the rather slow cISA (see Sect. 5.2) to implement the SI, i.e., even though the supporting atoms are unavailable. Therefore, there is no clear notion when an atom is *needed*, but rather when it may be *beneficial*.

Due to these differences, state-of-the-art LRU (least recently used)-based replacement policies may not be able to determine good replacement decisions for reconfigurable processors. Figure 4.26 shows the LRU behavior for the critical replacement decision when prefetching atoms for LF. When prefetching for LF starts, the atoms for ME are the least recently used ones. Therefore, according the LRU policy, they will be replaced first. Actually, this is a disadvantageous replacement decision, as these atoms are most likely going to be reconfigured soon again, because they lead to a noticeable performance improvement for the upcoming ME execution. The reason for LRU's disadvantageous replacement decision is the fact

---

[24]That is, the computations are completed before all demanded atoms are reconfigured.

[25]That is, already loaded to the reconfigurable fabric.

[26]Determined by SI forecasting (see Sect. 4.3).

[27]Determined by molecule selection (see Sect 4.4).

that the access locality of atoms is already considered by the prefetching and should no longer be an optimization goal for the replacement policy.

As the least recently used atoms have a high chance of being used again (considering an iterative execution of the computational blocks in an application), the atoms that were used most recently (MRU) might be good replacement candidates. MRU replaces those atoms that were used by the directly preceding computational block. For instance, Fig. 4.26 shows that the atoms for the discrete cosine transformation (DCT) and the Hadamard transformation (HT) are replaced when the prefetching for LF starts. However, MRU does not consider that two different computational blocks might share atoms, e.g., some atoms for HT are also demanded for the sum of absolute Hadamard-transformed differences (SATD) SI in ME. Again, the replacement decision is disadvantageous, as some beneficial atoms for the directly succeeding computational block are replaced. The new replacement policy – called MinDeg – overcomes the drawbacks from LRU and MRU in these situations, as shown in Sect. 4.6.2.

Many reconfigurable processors like Molen [VWG+04] or OneChip [JC99] consider LRU-based replacement policies. In addition, these processors allow a compiler-determined replacement decision [JC99, PBV07], which may be beneficial if good knowledge on the application control flow and the size of the available reconfigurable fabric is available at compile time. However, if the application control flow depends on the input data or the reconfigurable fabric has to be shared among multiple tasks (thus, the available fabric per task is not fixed), then a compiler-determined replacement policy cannot provide adaptivity and an LRU-based replacement might fail to determine good replacement decisions, as motivated in Fig. 4.26.

Compton [CLC+02] extends LRU to additionally consider the reconfiguration delay of reconfigurable accelerators. A *credit* value is assigned to each accelerator and it is set to the accelerator's size (e.g., considering look-up tables or gate equivalents) when it is reconfigured or demanded. Whenever an accelerator needs to be replaced, the accelerator with the smallest credit value is selected and the credit values of all other accelerators are decreased. Therefore, a larger accelerator has a higher chance to remain in the reconfigurable fabric even if it is not used for a certain time but it can eventually be replaced if it is not demanded for a longer time. However, the potential performance improvement of an accelerator is not considered, e.g., two similar-sized accelerators might have a noticeably different performance impact (depending on their availability when they are demanded) and even a rather small accelerator might lead to a bigger performance improvement than a rather large accelerator. Furthermore, relevant architectures (like presented in Sect. 2.2.4) consider similar-sized accelerators for practicability reasons, e.g., to achieve a regular partitioning of the reconfigurable fabric and to provide dedicated communication points between the reconfigurable and the nonreconfigurable fabric. Similar-sized accelerators result in similar reconfiguration delays for the accelerators and thus the approach degenerates to LRU.

Besides reconfigurable accelerators, some approaches considered implementing tasks or subtasks in reconfigurable hardware. Ahmadinia [ABK+04] used an LRU strategy to replace tasks from a task graph. Resano [RM04] proposed an adaptation of the Belady replacement [Bel66]. They exploit the property that their system can pre-determine a schedule for the subtask execution sequence, as they assume that the task-graph control flow and the subtask execution time are fixed. Therefore, knowledge about the near future of the subtasks execution sequence is available, which is then used to determine the replacement decision. However, this approach is only possible if the execution sequence of the subtasks can be predetermined and is not suitable for replacing accelerators, as the control flow is often input-data dependent and the execution sequence and execution time cannot be predetermined at compile time.

Instead of adapting existing replacement policies for cache and page replacement, limiting the scenario to predetermined execution sequences (to obtain future knowledge), or focusing on the reconfiguration delay, directly considering the performance impact of the reconfigurable atoms is required to fully exploit the potential of reconfigurable processors. Figure 4.27 shows an example for the SIs SATD and HT that were introduced in Fig. 4.26. An excerpt of the molecules is given along with the corresponding latencies of a molecule execution. The fastest currently available molecule of an SI is determined by the available atoms. For instance, when the atoms $\vec{a} = (0,2,1,1)$ are available (i.e., two instances of "sum of absolute values (SAV)", one instance of "Repack," and one instance of "Transform"), then the latencies of the fastest molecules in Fig. 4.27 for SATD, 4×4 HT and 2×2 HT are 93, 16, and 2 cycles, respectively. When the "Transform" atom would be replaced (leading to $\vec{a}' = (0,2,1,0)$) the latencies would slow down to 216, 174, and 67 cycles, respectively. Here, all three SIs are affected in a negative way because this atom is beneficial for all of them alike, i.e., it is shared among the SIs. Instead, replacing one of

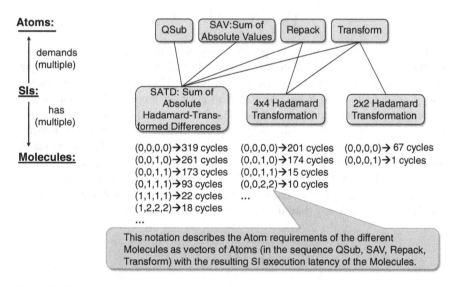

Fig. 4.27 Examples for atoms and their utilization in SIs, showing different implementation alternatives (i.e., molecules) and their execution latencies

the two SAV instances (leading to $\vec{a}'' = (0,1,1,1)$) would not affect the latencies at all. This is the key observation toward a performance-guided replacement, i.e., instead of considering the utilization history of the atoms (like LRU and MRU do), determining the performance impact for the SIs when replacing an atom.

4.6.2 The MinDeg Replacement Policy

Whenever a prefetching operation is about to start a new reconfiguration and no free space (i.e., not yet occupied parts of the reconfigurable fabric) is available, then an existing atom has to be replaced. At first, potential replacement candidates have to be determined out of which the replacement policy can choose one atom [BSH09b]. These candidates assure that no atom is replaced which is actually demanded by SIs of the computational block (e.g., LF) for which the prefetching started. The currently available atoms are denoted as $\vec{a}$ and the demanded atoms (determined by molecule selection, see Sect. 4.4) as $\vec{s}$. Some of the atoms that are demanded by $\vec{s}$ might be already available and shall not be replaced. The replacement candidates $\vec{c}$ are then given by Eq. 4.31. Altogether, the atoms in the meta-molecule $\vec{a} \triangleright \vec{s}$ will be reconfigured[28] and demand at most $|\vec{a} \triangleright \vec{s}|$ replacements. The general replacement function R (see Eq. 4.32) is specified to determine which atom A_i shall be replaced for the first reconfiguration (i.e., $R(1)$), the second reconfiguration (i.e., $R(2)$), and so on.

$$\vec{c} := \vec{a} \setminus \vec{s} \tag{4.31}$$

$$R : \left[1, |\vec{a} \triangleright \vec{s}|\right] \rightarrow \left\{A_0, \ldots, A_{n-1}\right\} \tag{4.32}$$

The input and output sets of this general replacement function R are now specified, but it is not yet defined *which* atoms (output of R) shall be replaced and *when* (input of R). This actually depends on the chosen policy. However, independent of the policy, all definitions of R have to assure an essential property. To formulate this property, the convenience function T from Eq. 3.10 is needed (see Sect. 3.4) to transform an atom into a molecule that comprises only that atom. The essential property of R that has to be true for all replacement policies is that the sum of all atoms that R selects for replacement has to match the initial meta-molecule of replacement candidates $\vec{c}$ (see Eq. 4.31). For instance, if $\vec{c}$ contains two instances of the atom A_i then there have to be exactly two different replacement times $x, y, x \neq y$ with $R(x) = R(y) = A_i$ and there must not be any third time $z, x \neq z \neq y$ with $R(z) = A_i$.

$$\sum_{i=1}^{|\vec{a} \triangleright \vec{c}|} T(R(i)) = \vec{c} \tag{4.33}$$

---

[28]If all planed reconfigurations finish before the next forecast.

The performance-guided minimum degradation (MinDeg) replacement policy is presented in detail [BSH09b]. To assure that it complies with the essential property from Eq. 4.33 the meta-molecule of replacement candidates is iteratively updated. This means that the algorithm starts with the initial replacement candidates in $\vec{c}$ (determined once after the prefetching is decided) and after each replacement decision, the replacement candidates are updated accordingly. In addition, not all replacement decisions are calculated in advance, but only on demand, i.e., when a new reconfiguration shall be performed. This is beneficial, when the computational block does not run long enough to perform all reconfigurations.

Despite of these general properties, the main difference of the proposed MinDeg replacement policy compared to the established cache/page replacement policies (as discussed) is its main replacement objective. Current state-of-the-art replacement policies [Tan07] are history based, i.e., they consider past events like "*when was it used,*" "*how often was it used,*" or "*when was it reconfigured*" (see Table 4.3). The MinDeg replacement policy is instead performance guided, i.e., for each replacement candidate (i.e., an atom) it examines the performance of all SIs after a potential replacement of that atom.

Algorithm 4.4 presents the pseudo code of the MinDeg replacement policy. The lines 4–16 correspond to the outer loop that examines all replacement candidates. The inner loop from line 7 to 11 iterates over all SIs and determines the latency of the fastest molecule for this SI assuming the current replacement candidate of the outer loop would actually be replaced. These latencies are accumulated in line 10 and in lines 12–15 the locally best replacement candidate (i.e., leading to the fastest accumulated latencies of all SIs) is memorized. After the replacement decision is determined, the list of candidates is updated in line 17 for the next replacement. Note, that the vector $\vec{a}$ of available atoms is not updated for the next replacement because it is also affected by the prefetching, i.e., the decision which atoms shall be reconfigured next. Therefore, the accurate information for $\vec{a}$ is

Table 4.3 Relevant history-based replacement policies, used for evaluating the performance-guided MinDeg policy

Policy	Description	Examined information
LRU	Least recently used	When was it used?
MRU	Most recently used	
LFU	Least frequently used	How often was it used?
MFU	Most frequently used	
FIFO	First in first out	When was it
LIFO	Last in first out	reconfigured?
Second Chance/ Clock	Extension of FIFO: Each atom in the queue has a flag that is set when it is used. When an atom shall be replaced (according to the FIFO policy) but the flag is set, it gets a second chance, i.e., its flag is cleared and it is moved to the end of the FIFO queue. "Clock" is a different implementation of the same policy	

expected as updated input for each replacement, whereas the replacement candidates are provided as input for the first replacement after a forecast and is then updated internally.

Algorithm 4.4 The performance-guided MinDeg replacement policy

1. // **Input:** available atoms $\vec{a}$ and replacement candidates $\vec{c}$
2. // **Output:** The atom A_r that is to be replaced next
3. $smallestOverallLatency := maxInt;$
4. $\forall A_i : T(A_i) \leq \vec{c}\{$
5. $\vec{t} := \vec{a} \setminus T(A_i);$ // $\vec{t}$ contains the remaining atoms, assuming A_i would be replaced
6. $overallLatency := 0;$
7. $\forall SpecialInstructionsSI_j\{$
8. // see Table 3.3 for the definition of functions
9. $SILatency := SI_j.getFastestAvailableMolecule(\vec{t}).getLatency();$
10. $overallLatency += SILatency;$
11. $\}$
12. if $(overallLatency < smallestOverallLatency)\{$
13. $smallestOverallLatency := overallLatency;$
14. $A_r := A_i;$
15. $\}$
16. $\}$
17. $\vec{c} := \vec{c} \setminus T(A_r);$ // Update for next replacement
18. return $A_r;$

The computational complexity of MinDeg is $\mathcal{O}(\#Atom-types \times \#SIs \times \#MoleculesPerSI)$. The term $\#MoleculesPerSI$ corresponds to line 9 in Algorithm 4.4 that determines the fastest implementation of an SI – considering a given meta-molecule $\vec{t}$ of atoms – by examining all molecules of that SI. However, a property from $\vec{t}$ can be exploited to reduce the amount of loop iterations. The vector $\vec{t}$ is created from the currently available atoms $\vec{a}$ by removing one of them. Therefore, $\vec{t}$ cannot provide a faster implementation of any SI SI_j than $\vec{a}$ does. Instead, the fastest available implementation in $\vec{a}$ might no longer be available in $\vec{t}$. Let us assume a list L_j of molecules from SI_j (sorted by increasing latencies) would be available with the constraint that all atoms demanded by these molecules are available in $\vec{a}$. Then, it is no longer required to iterate over all molecules of SI_j to determine the fastest one after a potential replacement. Instead, the first molecule $\vec{m}$ in L_j with $\vec{m} \leq \vec{t}$ is the fastest available molecule. L_j is initialized once when the application starts and then incrementally update it after each replacement and reconfiguration activity, i.e., further available molecules after a reconfiguration are inserted and no-longer-available molecules are removed after a replacement.

4.6.3 Evaluation and Results

The results of the performance-guided MinDeg replacement policy are analyzed. To determine the replacement quality, the execution time of an H.264 video encoder (like motivated in Sect. 3.3) is compared between MinDeg and state-of-the-art replacement policies. Due to its challenging computational requirements, the H.264 encoder demands multiple reconfigurations and replacements per video frame and thus clearly exposes the quality of the replacement policy. In addition, detailed insight into the actual replacement decisions is presented. Note that the MinDeg replacement policy is by no means specific to H.264. Instead, it aims to maintain a good overall SI performance and is thus generalized for any SI execution sequence rather than a specific one.

In addition to the initially motivated LRU/MRU example, also LFU/MFU, LIFO/ FIFO, and Second Chance (see Table 4.3) [Tan07] are benchmarked. Figure 4.28 shows the results of the comparison for reasonable reconfiguration bandwidths (a = 10 MB/s, b = 20 MB/s, and c = 40 MB/s) and different-sized reconfigurable fabrics (*x*-axis). The reconfiguration bandwidth determines the time to load new configuration data into the reconfigurable fabric. The performance penalty when replacing performance-wise *important* atoms is higher when the reconfiguration bandwidth is rather low. The reconfiguration bandwidth is limited by two major factors: (1) the bandwidth of the reconfiguration port and (2) the bandwidth at which the reconfiguration data can be streamed from memory to this port. The bandwidth of the internal configuration access port (ICAP, [Xil07b, Xil09c]) of the Xilinx Virtex Family has made noticeable improvements. The Virtex-II is specified to support 50 MB/s (8 bit at 50 MHz, but typically also working at a higher frequency) and for Virtex-4 it is extended to 400 MB/s (32 bit at 100 MHz). However, delivering the reconfiguration data at that speed implies certain drawbacks. On the one hand, the configuration data could be stored on a dedicated external memory (e.g., SRAM with 32-bit data port at 100 MHz), but this implies extra cost for a fast external memory and dedicated I/O pads. On the other hand, the data could be stored in the normal system memory that is also used to provide instruction- and data memory for the application. However, the reduced memory bandwidth that is available to the application during a reconfiguration might noticeably affect the application's execution time. Therefore, even though the configuration port is no longer the bottleneck in recent FPGA generations, a good performance for a restricted reconfiguration bandwidth is very important for overall system cost and performance.

The size of the reconfigurable fabric (*x*-axis in Fig. 4.28) determines how many atoms can be available at the same time. For a rather small reconfigurable fabric, typically all atoms will be replaced, thus alleviating the effects of a bad replacement decision. When more atom containers (ACs) are available, then more reconfigurations/replacements are performed and some atoms might not need to be replaced. For a rather large reconfigurable fabric, the amount of reconfigurations (and thus replacement decisions) might reduce; in the trivial case, all ever-demanded atoms will fit onto the reconfigurable fabric.

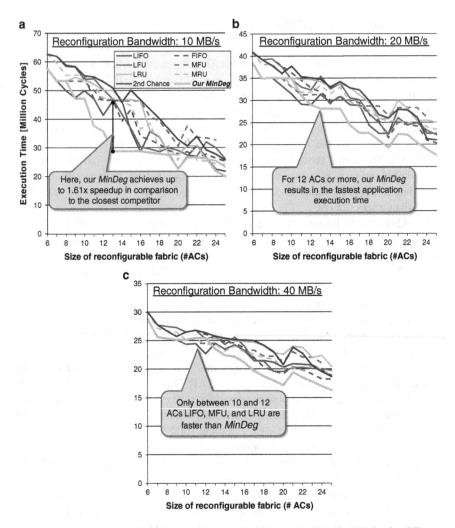

Fig. 4.28 Comparing the MinDeg replacement policy with state-of-the-art policies for different reconfiguration bandwidths (**a–c**) and size of the reconfigurable fabric (*x*-axis)

The comparison in Fig. 4.28 shows that MinDeg is superior in most of the scenarios. Especially for 20 and 40 MB/s reconfiguration bandwidths, it consistently leads to the fastest application execution for reconfigurable fabrics with at least 13 ACs. For 10 MB/s and 13 ACs, an overall application speedup of up to 1.61× in comparison to the closest competitor at that point (i.e., LIFO) is achieved. In all scenarios where MinDeg does not achieve the highest performance, its results are comparable or (in rare cases) down to 0.89× to the closest competitor.

To analyze the distribution of the performance results further, Fig. 4.29 presents box plots to summarize the speedup results for different reconfiguration bandwidths (5, 10, 15, …, 40 MB/s) and reconfigurable fabric sizes (5–25) in comparison to

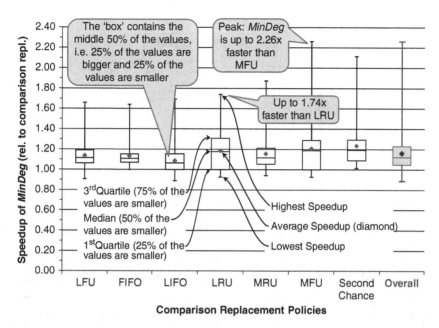

Fig. 4.29 Summarizing the performance improvement of MinDeg in comparison to state-of-the-art replacement policies

relevant replacement policies (see Table 4.3). The fact that the "boxes" (representing the middle 50% of the speedup values) are always above 1.0× speedup shows, that the MinDeg policy is generally beneficial. Actually, for 90.31% of all 1,176 benchmark comparisons, MinDeg reaches at least the same application performance and achieves up to 2.26× faster application execution (MFU, 5 MB/s, 15 ACs) in comparison to relevant replacement policies. In the remaining 9.69%, MinDeg never leads to an application performance less than 0.89× (LIFO, 15 MB/s, 10 ACs).

The fact that the boxes are relatively small (especially in comparison to the maximal and minimal speedup) shows, that MinDeg provides relatively stable performance improvements for a wide range of comparison scenarios. Altogether (summarizing the results of all 1,176 comparisons in the "Overall" box), MinDeg provides an application speedup between 1.05× and 1.23× for the middle 50% of all comparison benchmarks (in average 1.16× for all benchmarks). In comparison to LRU, MinDeg provides up to 1.74× speedup (10 MB/s, 13 ACs) and in average leads to a 1.18× faster application execution time.

Before evaluating implementation details and the overhead that is required to achieve these performance improvements, a detailed replacement analysis is presented, i.e., showing a specific run-time scenario and analyzing which atoms were replaced along with the resulting performance impact. In Fig. 4.30, the run-time details for the H.264 encoder with 20 MB/s reconfiguration bandwidth and a reconfigurable fabric that provides 15 ACs are shown. The execution details are compared for (a) the MinDeg policy, (b) LRU, and (c) MRU between 6 and 12 million cycles after the application started encoding (initial reconfigurations are performed

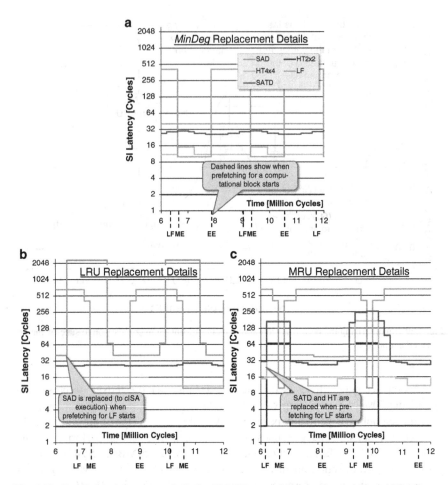

Fig. 4.30 Detailed replacement analysis for 20 MB/s reconfiguration bandwidth and 15 ACs

and further reconfigurations demand replacement). The *x*-axis furthermore shows when prefetching for a computational block (ME: motion estimation, EE: encoding engine, LF: loop filter) starts (the computational block itself starts right afterwards) and thus, the replacement policy is triggered. The *y*-axis shows the SI latencies in cycles on a logarithmic scale and the different lines in the figure correspond to different SIs. Thus, the replacement decisions and especially their performance impact are directly visible as the changing SI performance and the actual performance when an SI is demanded. For clarity, not all SIs were plotted. Instead, it focuses on the relevant SIs for ME (SAD and SATD), EE (HT2×2 and HT4×4), and LF to show the performance-wise relevant differences of the replacement policies. The omitted SIs for EE are also important for the overall execution time, but the replacement decisions for these SIs do not provide further insight to the differences between various replacement policies.

In Fig. 4.30a it can be seen that prefetching for LF starts at 6.35 million cycles. Right afterwards, an atom for SATD (and later also HT4×4) is replaced and affects the performance of these SIs. Therefore, the LF SI is accelerated (latency improves from 414 to 10 cycles per loop iteration of the SI) and right afterwards, the computational block completes. At 6.66 million cycles, prefetching for ME starts and the recently replaced atom for SATD (demanded by ME) are reconfigured again. It gives the impression that MinDeg should not have replaced the atoms for SATD a short time before they are demanded, but it is important to note that SATD was still in a fast hardware implementation (30 instead of 26 cycles per SI execution) and the performance of SAD (also demanded for ME) was not affected at all.

The major difference of the LRU replacement in Fig. 4.30b in comparison to MinDeg replacement in Fig. 4.31a is that the atoms for SAD are replaced such that this SI has to execute using the cISA (2,348 instead of 41 cycles per loop iteration) when ME starts. The reason is that the atoms for SAD have not been demanded for the longest time when LF prefetching starts. Some atoms for SATD were used in EE as well, so they are replaced later. The slow SAD performance and the relatively long time to achieve a hardware molecule again are the main reasons for the performance loss of LRU compared to MinDeg. The difference to the MRU policy in Fig. 4.30c becomes immediately visible when comparing the LF latency changes. At the beginning of ME prefetching, the previously used LF SI is replaced again. This is actually a good decision. However, the replacement decisions when prefetching for LF itself lead to the overall performance degradation compared to MinDeg replacement. The most recently used atoms at that time are those demanded

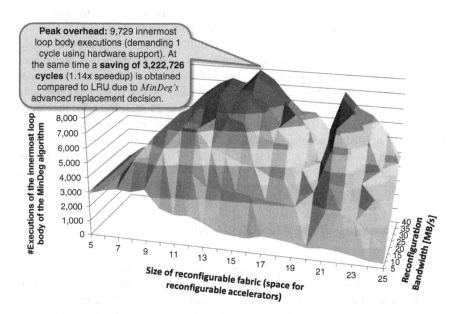

Fig. 4.31 Algorithm execution time (accumulated number of innermost loop body executions when encoding 10 frames)

by ME, especially the atom for the Hadamard transformation. Replacing these atoms leads to noticeable performance degradation for HT2×2, HT4×4, and SATD (169 instead of 31 cycles per SI execution), which affects the ME performance in a negative way. The MinDeg policy avoids replacement decisions that would lead to a noticeable performance degradation of any SI. Therefore, it maintains the relevant SIs SA(T)D and HT in a relatively fast implementation.

The previous sections have presented and analyzed the performance improvements that are achieved due to the proposed performance-guided MinDeg replacement policy. However, a certain overhead to execute the algorithm of the proposed replacement policy is required to obtain the application speedup. MinDeg's overhead and complexity are analyzed. The innermost loop body is in average executed 57.21 times per replacement decision. The innermost loop is inside the function to find the fastest available molecule, i.e., $SI_j.getFastestAvailableMolecule(\vec{t}).getLatency()$ (see line 9 in Algorithm 4.4). Each iteration of the $getFastestAvailablemolecule$ loop body (see Table 3.3) needs to determine whether a molecule (that implements SI_j) is smaller or equal than the meta-molecule $\vec{t}$ and the complexity of an iteration is bound by the number of atom types. With hardware support, such an inner loop iteration can be performed in a single cycle (using one comparator per atom type). The amount of inner loop executions depends on the computational block for which the reconfigurations are preformed (e.g., LF demands fewer atoms and, therefore, more replacement candidates are available). In addition, it depends on the size of the reconfigurable fabric. For instance, for 5 and 25 ACs, the innermost loop body is on average executed 31.22 and 62.13 times per replacement decision, respectively. However, the absolute introduced overhead does not only depend on the complexity per replacement decision, but it additionally depends on the amount of replacement decisions that are performed (depending upon the size of the reconfigurable fabric and the reconfiguration bandwidth). Figure 4.31 presents a surface plot that shows the absolute introduced overhead (i.e., accumulated over all replacement decisions) when encoding 10 frames with the H.264 video encoder application, depending on the reconfiguration bandwidth and the size of the reconfigurable fabric. The peak overhead is observed for 35 MB/s and 15 ACs. Here, the innermost loop body is executed 9,729 times altogether. This represents the demanded overhead that is needed to calculate MinDeg's advanced replacement decision, which – in this scenario–leads to a saving of 3,222,726 cycles (compared to LRU), corresponding to a 1.14× application speedup.

4.6.4 Summary of the Atom Replacement

Reconfiguring atoms at run time also demands replacing currently available atoms as the space of the reconfigurable fabric is limited. It was motivated and benchmarked, why state-of-the-art replacement policies (as used for cache and page replacement) are not necessarily beneficial for this purpose. The major difference is that cache and page replacement policies aim to exploit the access locality to cache lines or memory pages. However, for reconfigurable processors the access locality is already considered by the prefetching mechanism, because the reconfiguration time is rather long in comparison to loading a cache line or memory page. Therefore, the novel minimum degradation replacement policy MinDeg was developed that considers the potential performance degradation for SIs when replacing an atom and then replaces that atom that leads to the smallest performance degradation for the SIs. This policy does not rely on future knowledge or on analysis of previous execution patterns; instead, it aims to maintain all SIs in a good performance. It exploits the fact that modular SIs are upgradeable and correspondingly searches for the downgrade step with the smallest performance impact.

4.7 Summary of the RISPP Run-Time System

Using a run-time system to determine which atoms shall be reconfigured and when these reconfigurations shall be performed is essential to provide high adaptivity to changing situations and requirements and thus to exploit the full potential of run-time reconfigurable processors. In particular, it allows reacting on changing amount of reconfigurable fabric (e.g., due to multiple tasks sharing the reconfigurable fabric) and changing SI execution frequency (e.g., due to input data-based changing application control flow) adaptively. An online monitoring is used together with an error back propagation scheme to obtain a forecast of the upcoming SI requirements together with the expected execution frequencies of these SIs [BSTH07]. This information triggers RISPP's novel run-time system [BSH08b] to select a molecule for each SI that is forecasted for the upcoming computational block [BSH08c]. Afterwards, reconfiguring the atoms that are needed to implement these selected molecules is started. The highest efficiency first scheduler HEF determines the sequence in which these atoms shall be loaded by exploiting molecule upgrades and considering the predicted SI execution frequencies [BSKH08]. Whenever an atom shall be reconfigured, a currently existing atom may need to be replaced. The minimum degradation replacement policy MinDeg examines the performance impact for all replacement candidates and chooses the atom that leads to the smallest performance degradation for the SIs (i.e., the latency-wise smallest molecule downgrade) [BSH09b].

The novel algorithms for forecast fine-tuning, molecule selection, reconfiguration-sequence scheduling, and atom replacement are described on a formal basis and they are implemented and benchmarked. For forecast fine-tuning and molecule selection, different parameters exist that allow changing the fine-tuning behavior and the molecule profit function, respectively. The meaning of these parameters is described and their effect on the algorithms is benchmarked in the corresponding Sects. 4.3 and 4.4.4. For the reconfiguration-sequence scheduling, this monograph presents different algorithms and benchmarks them for different availability of atom containers in Sect. 4.5.3. The atom replacement policy was compared with state-of-the-art replacement policies (as they are used for cache and page replacement) in Sect. 4.6.3. In addition to benchmarking the algorithms, the RISPP approach is compared with state-of-the-art reconfigurable processors and nonreconfigurable ASIPs and the results are presented in Chap. 6. As the forecast fine-tuning is tightly coupled to the core pipeline and needs to decode SI and count SI executions, it is realized as a hardware implementation that is presented and evaluated in Sect. 4.3.3. The algorithms for molecule selection, reconfiguration-sequence scheduling, and atom replacement are triggered by forecast instructions and thus their coupling to the core pipeline is not that tight. Therefore, they may be implemented as hardware and/or software. For the hardware prototype, they are implemented as software, running on a Microblaze soft core, as shown in Sect. 5.5.

Chapter 5
RISPP Architecture Details

The novel modular special instructions (SIs) are described in Chap. 3 and the novel run-time system that dynamically determines the reconfiguration decisions to exploit the features of modular SIs is described in Chap. 4. To implement modular SIs, connect them to the core pipeline, and allow a run-time system to determine reconfiguration decision, a specialized processor architecture is needed to support these features, i.e., the RISPP architecture. In Sect. 4.1 and, in particular, in Fig. 4.1, a first overview of the RISPP architecture is given. There, it was already pointed out that it is not intended to define a completely new processor architecture. Instead, RISPP builds upon an existing architecture that is extended toward the RISPP requirements. In particular, in the scope of the presented work, a DLX core processor (created with ASIP Meister [ASI]) and later a SPARC V8 core processor (Leon2 [Aer]) were examined. This chapter focuses on the implementation details of the Leon2 prototype, although the general concepts are applicable to other architectures as well.

The first section describes how the instruction set architecture (ISA) of the core processor is extended to support SIs and which parameters they may obtain (e.g., registers, etc.) [BSH08a]. The second section describes how an SI can be executed using the core instruction set architecture (cISA) (when not all required atoms are available yet) [BSH08a] and Sect. 5.3 describes the data memory access and its realization that is provided for SIs. Section 5.4 presents the novel atom infrastructure that allows implementing modular SIs. It provides a computation and communication infrastructure that assures that (a) atoms can be reconfigured without affecting the computation of other components, (b) SIs can be upgraded by providing means to support rather small and rather large molecules, and (c) SIs may exploit the provided data memory bandwidth maximally [BSH08a]. Section 5.5 presents implementation results for the entire RISPP prototype, including the floorplan of the design, area requirements, frequency, and execution performance of the implemented run-time system. The RISPP prototype is implemented on an FPGA-based system shown in Appendix B.

L. Bauer and J. Henkel, *Run-time Adaptation for Reconfigurable Embedded Processors*, DOI 10.1007/978-1-4419-7412-9_5, © Springer Science+Business Media, LLC 2011

Fig. 5.1 Using dual-ported
BlockRAMs to implement a
general-purpose register file
(GPR) with one write and
four read ports

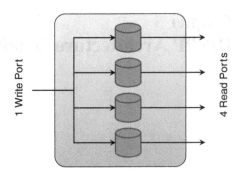

5.1 Special Instructions as Interface Between Hardware and Software

For processor architectures, the ISA is an interface between hardware and software. It describes a set of assembly instructions and defines their behavior without specifying their implementation. For instance, the ISA does not define whether the processor that implements it chooses an out-of-order superscalar implementation or an in-order pipeline for realization. Therefore, an application programmer (or compiler) can rely on the ISA without knowing all details of their implementation. The same is true for special instructions (SIs), which correspond to an instruction set extension. The concept of modular SIs (see Chap. 3) extends this concept by changing the implementation of the SIs during run time while maintaining the SI interface and the SI behavior to the application programmer (i.e., only the SI latency changes). This section describes the interface to the application programmer for the proposed SIs, e.g., which instruction formats were chosen for SIs and which parameters they may obtain as input and/or output. These decisions are fixed at design time and thus should be carefully considered.

As SIs aim to achieve better performance by exploiting parallelism, it is important to provide sufficient input data to the SI such that independent computations can be performed in parallel. Therefore, the general-purpose register file (GPR) is extended to provide four read ports and two store ports. In addition, a high-bandwidth data memory connection is provided for the SIs (see Sect. 4.1), however, that is independent of the SI ISA except the fact that the four input registers are used to determine which addresses in the data memory shall be read (see Sect. 5.4.2). As the RISPP prototype is based on a Xilinx Virtex FPGA,[1] the available BRAMs (on-chip memory blocks) are used to implement the GPR. Each BRAM is 2 KB[2] large, can be configured to support different aspect rations (e.g., 1×16 K, 8×2 K, 16×1 K, etc.), and provides two truly independent access ports (i.e., independent address lines, data lines, clock frequencies, and aspect ratios) [Xil08b].

---

[1] Early work was done on an Virtex-II 3000 and 6000, whereas the final prototype runs on an Virtex-4 LX 160.

[2] Actually, each byte is equipped with an extra parity bit that is user specific but cannot be initialized with the FPGA bitstream; thus, each BRAM provides 18 Kbit.

Figure 5.1 shows how four BRAMs are combined to realize a GPR with four read ports and one write port. For each BRAM, one port is a dedicated write port and each write access is written to all four BRAMs simultaneously, i.e., they always contain the same data. Therefore, the second ports of the four BRAMs can be used to provide four independent read ports. However, this concept cannot be used to provide a second write port. Instead, a small finite state machine (FSM) is added to the write-back stage of the core processor that stalls the stage for one cycle and performs both write accesses sequentially. Even though it seems that this concept does not provide any performance improvement, it allows providing SIs that demand two write-backs. For instance, the computation of division and modulo of two numbers (i.e., both shall be computed together in one assembly instruction) is typically a multicycle instruction (i.e., the execute stage of the core processor is stalled until the computation completed) and creates two results. Without a second write port, two independent instructions need to be executed (one for division and one for modulo). As both instructions would stall the core pipeline, the concept of providing two (sequential) write ports actually leads to a performance improvement (if the execute stage is stalled for more than one cycle per instruction) as now only one instruction stalls the execute stage.

The SIs may now use the extended GPR. Therefore, they have to provide up to six potentially different addresses (four read and two write). As the Leon2 provides 32 registers,[3] 5 bits are needed to address a particular register. Addressing six registers with 5 bits each would demand 30 bits. As the SPARC ISA uses 32-bit instructions, only 2 bits would be left for further information, e.g., that the instruction corresponds to an SI and which particular SI it denotes. This would allow distinguishing at most three different SIs and therefore it is clear, that not all six registers can be addressed independently. As the other extreme, no register might be encoded explicitly, but the SI opcode might determine the demanded registers implicitly. For instance, Chimaera [HFHK97, YMHB00] allows reading nine registers for an SI execution, but for a given SI it is predetermined, which registers will be read. The compiler has to provide the demanded SI input in these registers, i.e., the hardware is rather inflexible here, and the compiler has to compensate this irregularity. In this work, a compromise is chosen. Up to four different registers shall be addressable, whereas any of these four may be read (utilizing all available read ports) and up to two of the same registers may be written. If an SI only demands that two registers are read and two are written, then all registers may be addressed independently. However, if an SI demands four input registers and two output registers, then the outputs will actually overwrite two of the input registers, as they share the addresses. In these cases, there is still an irregularity in the extended ISA for SIs, i.e., the compiler has to create copies for the input registers if their lifetime exceeds the point in time of the SI execution. In addition, it should be possible to provide immediate values to the SIs, e.g., a flag or a small constant.[4]

---

[3] Actually, the Leon 2 implements register windows, however, at most 32 registers are visible (and thus addressable) at the same time.

[4] Constant during compile time and run time but not during SI design time (otherwise it could be hard coded within the SI).

Table 5.1 SPARC V8 instruction formats

Format 1 (op=1): Call

op	disp30

bit# 31 30 29 28 27 26 25 24 23 22 21 20 19 18 17 16 15 14 13 12 11 10 9 8 7 6 5 4 3 2 1 0

Format 2 (op=0): SETHI & Branches

op		rd	op2	imm22
op	a	cond	op2	disp22

bit# 31 30 29 28 27 26 25 24 23 22 21 20 19 18 17 16 15 14 13 12 11 10 9 8 7 6 5 4 3 2 1 0

Format 3 (op=2 or 3): Remaining Instructions

op	rd	op3	rs1	i=0	asi	rs2
op	rd	op3	rs1	i=1	simm13	
op	rd	op3	rs1	opf		rs2

bit# 31 30 29 28 27 26 25 24 23 22 21 20 19 18 17 16 15 14 13 12 11 10 9 8 7 6 5 4 3 2 1 0

Table 5.2 SPARC V8 instruction format 2 with RISPP extensions

Value for $op2$	SPARC V8 allocation	RISPP extension
0	UNIMP	HI instruction group
1	Unused	SI without register write-back
2	Branch on integer condition code	
3	Unused	SI with one register write-back
4	Set register high 22 bits; NOP	
5	Unused	SI with two register write-back
6	Branch on floating-point condition code	
7	Branch on coprocessor condition code	

Table 5.1 shows the different instructions formats for the SPARC core ISA [SPA]. The field op determines one of the three instruction formats. Formats 2 and 3 are subdivided further, i.e., they provide the field $op2$ and $op3$, respectively. Table 5.2 shows the usage of different $op2$ values for Format 2 in Table 5.1. It contains three unused values (1, 3, and 5) that can be used to extend the ISA to provide SIs. In addition, the value 0 is only used to explicitly trigger an "unimplemented instruction" trap (UNIMP). Due to these freely available opcode values for $op2$, Format 2 is used for RISPP extensions of the SPARC ISA. As discussed beforehand, up to two registers may be written back to the GPR. The three "unused" values in $op2$ are used to encode the three different possibilities for SIs, i.e., writing two registers, one register, or no register at all. This determines the interpretation of the registers that are explicitly addressed in the remaining bits of Format 2 as is shown later. In addition to the SIs, also so-called helper instructions (HIs) are needed to support the RISPP concept [e.g., for forecast instructions (FIs), see Sect. 4.3]. They are packed into the $op2 = 0$ field that is also used to trigger the UNIMP trap. To assure that the normal UNIMP functionally is still operational, one specific opcode value is reserved for it. Table 5.3 shows how the bits 25–29 are used to distinguish between the UNIMP trap and different HI opcodes.

Table 5.3 SPARC V8 format 2 used for UNIMP and helper instructions (HIs)

SPARC V8 UNIMP format/opcode

op	reserved	op2	imm22
0 0	0 0 0 0 0	0 0 0	imm22

bit# 31 30 | 29 28 27 26 25 24 23 22 | 21 20 19 18 17 16 15 14 13 12 11 10 9 8 7 6 5 4 3 2 1 0

RISPP HI format/opcode

op	hi_op	op2	HI-specific information
0 0	hi_op	0 0 0	HI-specific information

bit# 31 30 | 29 28 27 26 25 24 23 22 | 21 20 19 18 17 16 15 14 13 12 11 10 9 8 7 6 5 4 3 2 1 0

Table 5.4 Overview of implemented helper instructions

for cISA	op		op2				
✓	0 0	REGMOV1	0 0 0	unused		rs4	rs2
✓	0 0	REGMOV2	0 0 0	unused		rs4	rs2
✓	0 0	REGSAV	0 0 0	unused	rs5	rs4	rs2
✓	0 0	SIID	0 0 0	unused			rs2
✓	0 0	IMOV10	0 0 0	unused			rs2
✓	0 0	IMOV5	0 0 0	unused		rs4	rs2
	0 0	RMVAR	0 0 0	unused		rs4	imm5
(✓)	0 0	SIV	0 0 0	unused			rs2
	0 0	SIGRP	0 0 0	unused			imm5
	0 0	FI	0 0 0	imm22			

bit# 31 30 | 29 28 27 | 26 25 | 24 23 22 | 21 20 19 18 17 16 15 14 13 12 11 10 | 9 8 7 6 5 | 4 3 2 1 0

The HI-specific information shown in Table 5.3 depends on the HI that the hi_op field specifies. Table 5.4 provides an overview of the HIs that are implemented to support the RISPP concept. Most of them accelerate the SI execution with the cISA and are explained in Sect. 5.2. The HIs of RMVAR, SIGRP, and FI are explained as follows.

RMVAR. Some parts of the run-time system may be configured by parameters. For instance, the forecast fine-tuning (see Sect. 4.3) has two parameters (α and γ) to configure the strength of the error back propagation. In addition, the molecule selection (see Sect. 4.4) has two parameters. This HI allows the user application (or an OS service) to adapt these parameters, i.e., it allows setting a variable in the run-time system (RMVar).[5] The imm5 value (see Table 5.4) specifies which parameter shall be set, and the rs4 register contains the new value of this variable.

SIGRP. As shown in Table 5.5, 5 bits are reserved in the SI instruction format to define the SI opcode. This allows distinguishing 32 different SIs. Even though this amount of SIs is expected to be sufficient for one application, more SIs will be demanded when multiple application shall be supported/accelerated. Therefore, a 10-bit SI opcode is used to be able to provide up to 1,024 different SIs. However, only 5 bits of it are determined in the instruction word. The remaining 5 bits are realized as a dedicated register in the decode stage that is concatenated with the

---

[5] Note, "RM" stands for rotation manager, indicating RISPP's run-time system.

Table 5.5 Instruction format for special instructions as part of SPARC V8 instruction format 2

op	rd	op2	imm	si_op	rs5	rs4	rs2
0 0	rd	op2	0 0	si_op	rs5	rs4	rs2
0 0	rd	op2	0 1	si_op	rs5	rs4	imm5
0 0	rd	op2	1 0	si_op	rs5	imm5	imm5
0 0	rd	op2	1 1	si_op	rs5	imm10	

bit# 31 30 | 29 28 27 26 25 | 24 23 22 | 21 20 | 19 18 17 16 15 14 13 12 11 10 | 9 8 7 6 5 | 4 3 2 1 0

op2 determines the register write back
001 : no write back
011 : rd write back
001 : rd & rs5 write back

imm determines the input immediate
00 : no immediates
01 : rs2 used as 5-bit immediate
10 : rs2 and rs4 used as two 5-bit immediates
11 : rs2 and rs4 used as 10-bit immediate

5 bits in the instruction word to derive the 10 bits of the SI opcode. This concept groups the SIs to clusters of 32 different SIs that share the same value of the dedicated register. This way, every application may use its own SI group that is changed as a part of the context switch. This HI allows setting the 5 bits in the dedicated register, i.e., to change the SI group.

FI. This HI corresponds to the forecast instruction (FI) that is used to predict the SI execution frequency of an upcoming computational block (see Sect. 4.3). The actual information is embedded into the imm22 field and is not further evaluated by the decode stage of the core pipeline. Instead, this information is forwarded to the hardware for fine-tuning the SI execution frequencies that extracts the information, which forecast block is addressed, i.e., start address and length are embedded into the immediate.

Let us now look at the instruction formats for the SIs. As discussed above, up to four registers need to be addressed for reading (in some cases, immediate values are preferable), up to two registers for writing, and 32 different SI opcodes need to be distinguished within the instruction word (leading to 1,024 different SIs in combination with the dedicated register for the SI group). Table 5.5 shows the different instruction formats for SIs. They use Format 2 of the SPARC V8 instruction formats (see Table 5.1), i.e., the value for the *op* field is always "00." The op2 field determines how many registers shall be written back (see Table 5.2) and thereby implicitly determines which fields address them (rd and/or rs5). Independent of the written registers, the registers that are addressed by *rd* and *rs5* are always read. In addition, the registers that are addressed by the fields *rs2* and *rs4* may be read or the values of these fields are used to compose immediate values (determined by the *imm* field). Eventually, the *si_op* field determines which SI – within the currently active SI group – shall be executed. For instance, if an SI demands two register inputs and creates two registers as output, then the combination *op2* = "001" and *imm* = "00" configures rs2 and rs4 as inputs and rd and rs5 as outputs. In this example, all four registers can be addressed independently. However, if further input data is demanded (e.g., immediate values or more registers), then the registers addressed by rs5 and rd are also read (in addition to rs2 and rs4 that are used as registers or immediate values) and will potentially be overwritten by the SI.

5.2 Executing Special Instructions Using the Core Instruction Set Architecture

In the scope of reconfigurable processors, it may happen that an SI shall execute but the hardware implementation of that SI is not yet reconfigured. Many early reconfigurable processor projects (e.g., [HFHK97, LTC+03, RS94, WC96, WH95]) stalled the core processor until the corresponding reconfiguration completed. This reduces the potential performance in comparison to architectures (e.g., [Dal03]) that provide an SI implementation using the cISA as demonstrated in Fig. 3.1. In addition, architectures without a cISA alternative for SIs may face the problem of configuration thrashing [Dal99]. For instance, consider a reconfigurable processor with a reconfigurable fabric that provides space for one SI implementation. If an application demands two SIs in an alternating way in an inner loop (i.e., SI_1, SI_2, SI_1, SI_2, ...), then each execution of SI_1 replaces the SI implementation of SI_2 (and vice versa) such that for each SI execution the corresponding implementation has to be reconfigured first. This leads to a significantly reduced performance, even when comparing with a conventional software implementation (i.e., without SIs and hardware acceleration). In this example, the fastest application execution demands changing the application to use only one SI (either SI_1 or SI_2, depending on their execution frequencies, etc.). However, in a multitasking environment this is not feasible, as it is not known at application compile time, how many reconfigurable fabric is available for a particular application (i.e., how many SIs can be loaded into the reconfigurable fabric in parallel). In an extreme case, an application might not obtain any reconfigurable fabric and thus would need to stall upon an SI execution until some reconfigurable fabric is assigned to it. The concept of executing an SI using the cISA conceptually solves these problems and the remainder of this section presents how it is realized in the scope of RISPP.

The concept of executing an instruction using the cISA is not specific to reconfigurable processors. For instance, many ISAs contain instructions for multiplication, division, or floating point operations that are not implemented on all processors that aim to implement that ISA. Therefore, they often provide a synchronous trap (typically called "unimplemented instruction") that is automatically triggered if the application executes an instruction that a particular processor does not implement in hardware. This mechanism is used to realize the SI cISA implementation. A dedicated memory is provided that contains for each SI the information, which implementation is the fastest one that is currently available. This memory is updated by the run-time system when the fastest implementation of an SI changes. Whenever the application executes an SI, the decode stage consider the information in this memory to determine whether an "unimplemented instruction" trap has to be triggered or not.

Figure 5.2 provides an overview of the different levels of the cISA implementation of an SI. After the trap was triggered, the execution continues at the trap table (identified through the trap base registers) that provides space for four instructions per trap. The trap base register together with the trap type determine the exact position in the trap table. From the trap table, the application jumps to the trap-type specific trap handler that is shown afterwards. For the cISA implementation, the

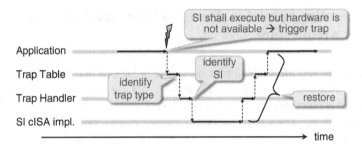

Fig. 5.2 Levels and contexts to reach the cISA implementation of an SI

trap handler has to identify which SI shall be executed, which parameters were provided to this SI, and to which registers the results shall be written. Afterwards it calls an SI-specific cISA implementation that calculates the result of the SI for the provided input data and returns the computed results. The trap handler writes these results to the expected registers and afterwards the execution continues with the instruction that follows the SI that caused the trap. In particular, the following operations are performed:

1. Jump to the trap table.
2. Jump to the "unimplemented instruction" handler code.
3. Save the processor state.
4. Identify the instruction that caused the trap.
5. Acquire source data.
6. Jump to the code that executes the identified SI.
7. Calculate result(s) of the SI.
8. Write-back result(s).
9. Restore the processor state and return.

Steps 4, 5, and 8 imply a significant performance overhead when executed with the cISA, as they heavily rely on accessing and parsing the 32-bit instruction word of the SI that caused the trap. For instance, to identify which SI caused that trap, the SI instruction needs to be loaded from instruction memory to a register (using the trap return address that points to the instruction after the SI that triggered the trap) and then this SI needs to be decoded by software to extract the SI opcode and register addresses. On the contrary, these steps correspond to simple operations when implemented with dedicated hardware. Therefore, helper instructions are added (HIs, see Sect. 5.1) to accelerate these parts. Step 4 applies a mask and a constant shift to extract the SI opcode. In hardware, this corresponds to a simple rewiring (i.e., extracting the bits of the SI opcode) and the results are saved in a dedicated register. This register can be read using the SIID HI (i.e., this HI copies the value of the special register into a GPR). Step 5 accesses the data from the register file and the immediate values that are provided from the decode stage. This data is also temporarily stored in dedicated registers when an SI executes and afterwards it can be read using the REGMOV1, REGMOV2, IMOV5, and IMOV10 HIs (see Table 5.4). REGMOV1 copies the content of the SI input parameters rd and rs2 (i.e., the content of the addressed registers) to the specified GPRs (using the fact that the extended GPR has two write ports). Similarly, REGMOV2 copies rs4/rs5, IMOV5 copies the two 5-bit immediate values, and IMOV10 the 10-bit immediate value. Step 8 copies the calculated results of the SI to its final destination registers. The information to which registers the results shall be written is extracted from the decode stage and temporarily stored. After the actual cISA implementation completed, this information is used to write the created results to these registers, using the REGSAV HI. This HI receives up to two results (rs2 and rs4) as input, and the content of the register that is addressed by rs5 decides whether one or two results shall actually be written back.

Algorithm 5.1 Trap handler to implement special instructions with the cISA and the support of the helper instructions

```
 1.  void unimp_handler() {
 2.      int si_id, regsav, g1, psr, rd1, rd2;
 3.      int rs1, rs2, rs4, rs5, imm10, imm5_1, imm5_2;
 4.      asm( "mov %g1, g1"  // save %g1 register
 5.           "mov %psr, psr"  // save CPU status register
 6.           "siid si_id"  // load SI identifier
 7.           "regmov1 rs1, rs2"  // load input registers rs1 and rs2
 8.           "regmov2 rs4, rs5"  // load input registers rs3 and rs4
 9.           "imov5  imm5_1, imm5_2"  // load the 5-bit immediates
10.           "imov10 imm10"  // load the 10-bit immediate value
11.      );
12.      switch (si_id)          { // jump to cISA execution
13.      case 0x2A:              // one showcase SI opcode
14.          regsav = 1;        // set amount of write-backs
15.          rd1 = ...          // here comes cISA execution
16.          break;
17.      case ...
18.          break;
19.      default:
20.          regsav = 0;        // set amount of write-backs
21.          break;
22.      }
23.      asm   ( "mov psr, %psr"  // restore CPU status register
24.           "mov g1, %g1"   // restore %g1 register
25.           "nop"
26.           "regsav rd1, rd2, regsave"  // SI register WB
27.           "restore"            // restore register window
28.           "jmpl %l2, %g0"  // set jump target (the instr. after the SI)
29.           "rett %l2 + 0x4"  // and return from handler
30.      );
31.  }
```

Algorithm 5.1 shows the resulting code for the trap handler, using optimized inline-assembly code to execute the HIs. For simplicity, the inline-assembly part directly accesses the C-Code variables by stating their names instead of declaring specific input/output names, as the actual GCC inline-assembly syntax demands it [Ric]. The switch-case statement in lines 12–22 is a placeholder for the actual cISA implementation. In addition to the SI computation, it sets the amount of register write-backs by means of the regsave variable. In total, an overhead of 38–39 cycles (depending on the number of register write-backs, see Sect. 5.1) is applied to the actual SI cISA execution with. For instance, SATD in Fig. 3.3 requires 319 cycles plus 38 cycles for the trap overhead. This corresponds to an 11.9% increased SI execution time for the cISA implementation, which is acceptable according the above-discussed conceptual problems that this approach solves. However, other implementation alternatives exist that also realize a cISA implementation. The advantage of the trap-handler methods is that the application programmer does not have to consider which implementation is used for an SI when writing it to the application code. Alternatively, the application programmer could explicitly write a conditional branch in front of each SI as indicated in Algorithm 5.2.

Algorithm 5.2 Examples for implicit and explicit cISA execution of SIs

```
1.  // implicit cISA execution if the SI HW is unavailable (using trap)
2.  int a = 42, b = 23, result; // prepare SI input registers
3.  asm("mySI result, a, b");
4.  // explicit cISA execution if the SI hardware is unavailable (using condi-
    tional branch)
5.  int x = 42, y = 23;
6.  if (is_hardware_available(MY_SI)) {// MY_SI is the constant SI opcode
7.  asm("mySI result, x, y");
8.  } else {
9.  result = ...// here comes cISA execution
10. }
```

The explicit cISA execution in Algorithm 5.2 has certain advantages and disadvantages. One disadvantage is that the application programmer has to encapsulate each SI manually with a conditional branch, as shown in the example. In addition, this conditional branch reduces the SI execution performance in the case that the SI hardware implementation is available because the condition needs to be calculated and a branch needs to be performed. However, in case that SI implementation is not available, it increases the performance in comparison to the implicit (i.e., trap-based) cISA execution. For instance, the overhead of the trap and extracting the SI opcode and parameters are not further required. In addition, the compiler might perform global optimizations on the code, for instance, constant propagation may be applied in the example in Algorithm 5.2 for the parameters x and y.

Instead, the constants for parameters a and b (for the implicit cISA execution) need to be loaded to registers to pass them to the SI, which – if the trap handler executes it – are extracted again and copied to other registers. Summarizing, the implicit cISA execution allows for an overhead-free implementation of the SI if the hardware implementation is available at the cost of 38–39 cycles trap overhead if the cISA implementation shall be used. The explicit cISA implementation minimizes the cISA overhead at the cost of a reduced SI performance if the hardware implementation is available. As the cISA implementation can be considered as an interim implementation, whereas the majority of the SI executions should use a hardware implementation, the implicit cISA execution is chosen. However, the explicit cISA execution or combinations of both are also feasible.

If a trap for a certain SI is triggered, it means that not all required atoms are available in hardware. Still, some of the atoms might be available. For instance, the SATD SI (see Fig. 3.3) is composed of multiple instances of four different atom types, i.e., QSub, Repack, Transform, and SAV. Even if not all atoms are available in hardware with at least one instance, it might be beneficial to use some of the available atoms and to compute the remaining parts using the cISA. If the QSub, Repack, and Transform atom are available with at least one instance, then their functionality might be used to accelerate the SI execution even though not all demanded atoms are available for a full hardware implementation. The access to a single atom is realized as a so-called *elementary* SI, i.e., it corresponds to an SI with two inputs and two outputs that are directly connected to the inputs and outputs of the atom. This implementation corresponds to the so-called mixed molecules as introduced in Sect. 3.2. At most one atom is used at a time, thus the molecule-level parallelism is not used. Still, these mixed molecules provide a performance improvement in comparison to the pure cISA execution, as the atom-level parallelism is exploited.

As some parts of a mixed molecule demand the cISA, also the "unimplemented instruction" trap is used to trigger their execution. Within the trap handler, software might be used that reflects the data flow of the SI and decides locally for each atom whether it shall execute using the cISA or whether it may use an elementary SI, i.e., whether that atom is available in the reconfigurable fabric. However, this leads to potential performance problems, as shown below. Instead of this, the trap handler for a certain SI probes once, which atoms are available, by using the SIV helper instruction (see Table 5.4) to determine the "SI version," i.e., which mixed molecule shall execute the SI. For each mixed molecule, a dedicated implementation is provided as part of the trap handler. The advantage of offering multiple SI implementations inside the trap handler (one for each mixed molecule and one for the pure cISA molecule) is that, for each implementation, the compiler statically knows which parts use the cISA (i.e., which atoms are not available) and can optimize them correspondingly. For instance, in the SATD example in Fig. 3.3, QSub and Repack execute subsequently and Repack obtains the result from QSub as input. If QSub and Repack would both be executed using the cISA, then the compiler may perform common subexpression elimination for the combined QSub/Repack software. Alternatively, just one code that is used for all mixed molecules and the cISA molecule together might be provided. In such a case, it needs to be determined within that software, whether an atom is available or not, using a conditional branch that executes either the atom or the cISA code that corresponds to the atom functionality.

In this example, if QSub and Repack would both execute using the cISA, then still the two conditional branches would encapsulate the cISA implementations of QSub and Repack, respectively. As the condition for the branches is not known at compile time, the compiler cannot optimize the code across the boundaries of atoms. In addition, not necessarily all combinations of atom availabilities are beneficial. For instance, if Repack is available but Transform is not, then it is typically not beneficial to use the Repack atom, because Repack is used in this SI to organize the data in such a way that it is suitable for the Transform atom. If the Transform atom is not available, then this data reorganization is not necessarily required.

5.3 Data Memory Access for Special Instructions

To exploit the performance potential of hardware-accelerated calculations by atom-level parallelism and molecule-level parallelism (see Sect. 3.2), the SIs need sufficient input data. For instance, the SATD SI (see Fig. 3.3) requires eight 32-bit inputs, some other SIs like SAD require even more. However, the instruction format for SIs and the register file only provide up to four different 32-bit registers as input (see Sect. 5.1). Therefore, additional data memory access is provided for the SIs, as indicated in Fig. 4.1. Obviously, a larger memory bandwidth increases the potentially available parallelism that may be exploited by SIs. However, it also increases the costs of the system. As a compromise, exactly the same memory connection is provided as Tensilica [Tena] offers them for their Xtensa LX2 ASIP [Tenb], i.e., two independent ports with 128 bits (quadword) each. Similar to Tensilica, the RISPP prototype uses on-chip memory to realize this data bandwidth and thus the memory capacity is limited. Using off-chip memory would imply a rather large amount of I/O pads to connect both ports to the chip.

The on-chip memory corresponds to a scratchpad memory that is shared between the core pipeline and the reconfigurable fabric because SIs may execute either on the core pipeline (cISA execution, see Sect. 5.2) or on the reconfigurable fabric (in case sufficient atoms are available) and need to be able to access the memory. As it is typical for shared memory, potential memory inconsistency problems may occur if an SI that accesses the memory executes on the reconfigurable fabric and – during the multicycle SI execution – load/store instructions execute in the core pipeline. For instance, the pipeline may load a value from memory, modify it, and write it back. If an SI on the reconfigurable fabric writes the same address in between, then the core pipeline overwrites its result. In the scope of the OneChip98 project [JC99], Carillo identified nine different memory inconsistency problems for which hardware support was developed to resolve them [CC01]. Only one application (JPEG encode/decode) of their four benchmarks actually took advantage from the ability to execute the core pipeline in parallel to the SIs. As the performance improvement was rather small (1.72%), Carillo concluded that "simply making the

CPU stall when any memory access occurs while the RFU[6] is executing, will not degrade performance significantly on the types of benchmarks studied" [CC01]. This rather small performance improvement is not surprising. The SIs are designed to cover the most prominent computational blocks of the application in parallel hardware. In addition, executing the computationally less important parts on the core pipeline in parallel to the SI execution should not affect the overall performance significantly. Based on these general considerations and on Carillo's investigation, the core pipeline is stalled during the SI execution. However, the proposed concept of modular SI does not imply a conceptual necessity to stall the pipeline. If it appears to be promising for any particular application scenario, the prototype may be extended accordingly.

Even though the core pipeline is stalled during an SI execution, there is still one potential memory inconsistency problem left. If a load/store instruction precedes an SI (i.e., it enters the pipeline one cycle earlier), it reaches the memory stage when the SI reaches the execute stage of the pipeline. If the SI immediately issues a memory access, then the access from the load/store instruction and the SI are issued in the same cycle. However, solving this problem is rather simple. As the load/store instruction was issued before the SI, it should also be allowed to access the memory before the SI. Therefore, an arbiter that schedules the memory accesses from the memory stage and the SIs is provided and it gives priority to the memory stage statically.

As two different types of memory (on-chip cache and on-chip scratchpad) with different ports are available and as additionally two different processing resources (core pipeline and reconfigurable fabric for SI implementation) are available that access both memory types, a memory controller is developed to connect them. Figure 5.3 gives an overview, how the memory controller is connected to the system.

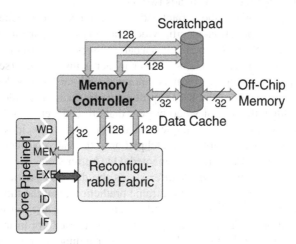

Fig. 5.3 Memory controller, connecting the memory stage of the core pipeline and both 128-bit ports for SIs with the data cache and an on-chip scratchpad memory

---

[6] Reconfigurable functional unit (RFU) corresponds to a reconfigurable region into which an SI implementation may be loaded.

The core pipeline as a 32-bit port from the memory stage and the reconfigurable fabric has two independent 128-bit ports. The cache has a 32-bit port while the scratchpad has two 128-bit ports. The core pipeline may also access the scratchpad, using 32-bit data. The scratchpad provides byte-enable signals that allow writing the particular 32-bit word (or even a single byte). The core pipeline may access words (i.e., 4 bytes) at addresses that are word aligned (i.e., the address is dividable by four without remainder). However, these addresses are typically not quadword aligned as demanded by the scratchpad. Therefore, the memory controller creates the corresponding quadword address and provides the 32-bit data at the expected bit positions. If an SI accesses the data cache from the reconfigurable fabric, then the memory controller serializes the accesses. The SI may request up to 256 bits at the same time, which corresponds to eight different word accesses. If an SI performs a 128-bit access to the scratchpad that is not quadword aligned, then the memory controller automatically creates up to two subsequent quadword accesses to realize them. For instance, if a quadword access to address 0b000010[7] is requested (each byte has an individual address), then this address is not quadword aligned (the last 4 bits are not zero) and just cannot be directly passed to the memory. Instead, the memory controller issues two accesses at addresses 0b000000 and 0b010000. Each one loads 16 bytes and the memory controller combines these 32 bytes to return the requested 16 bytes from address 0b000010 to 0b010001. However, if afterwards the subsequent 16 bytes are requested (i.e., 0b010010) then the memory controller only issues one additional access to address 0b100000 and combines that data with the 16 byte from the last request, i.e., the last request is buffered. Therefore, in a sequence of 128-bit accesses that are not quadword aligned, only the first request demands two memory accesses.

In the FPGA prototype, the scratchpad is partitioned into a general-purpose scratchpad and an application-specific video memory. This partitioning is not a part of the general RISPP concept, but it is a part of the periphery of the surrounding system. For benchmarking, an H.264 video encoder is investigated (see Sect. 3.3) and the FPGA prototyping platform is equipped with an audio/video periphery module that is connected to the RISPP processor. To access the video input data with a high bandwidth, special video buffers that are connected to the prototype like the scratchpad memory are implemented. The address decides whether an access goes to the general-purpose scratchpad or to the video buffers. Altogether, three different video buffers are implemented. One buffer stores the current frame that shall be encoded, the second buffer stores the decoded previous frame (as demanded by video encoders with intraframe prediction), and the third one stores the next frame that is read from the video camera during the current frame is encoded. In addition, the general-purpose scratchpad is used, for instance, to store the temporary results of the DCT calculation. For encoding, the motion estimation compares the current frame with the decoded version of the previous frame

---

[7]For simplicity, only six bit addresses are written and the last four bits (indicating which byte of a quad-word shall be accessed) are underlined.

(SAD and SATD SIs) accessing both video buffers with one 128-bit port, respectively. After encoding, the frame is decoded to serve as a comparison frame for the next frame. This decoded data overwrites the current frame, using both ports on that buffer. One port of the buffer that contains the previous frame is connected to a VGA output periphery module. This is possible, as no SI demands accessing the previous frame with both ports. After encoding and decoding are completed and a new frame is available from the video camera, the buffers rotate their role, i.e., the input frame from the camera becomes the current frame, the current frame becomes the previous frame, and the next frame that comes from the camera over-writes the old previous frame. This rotation is performed in hardware by changing the connections of the buffers. This simplifies the software development, as the current frame can always be accessed at one particular address, independent of which of the three video buffers currently contains the current frame. The video core that performs this buffer rotation is memory-mapped to the processor to control when such a rotation shall take place.

5.4 Atom Infrastructure

The previous section described how the SIs are embedded into the instruction format of the core pipeline, how they are executed using the cISA, and how they maintain sufficient input data. However, to execute an SI using the reconfigurable atoms, a special infrastructure – the so-called atom infrastructure – is demanded that fulfills the following requirements:

1. Offer small reconfigurable regions (so-called atom containers, ACs) such that each of them can be reconfigured to accommodate one atom at a time, without affecting the other regions during reconfiguration.
2. Provide a communication infrastructure that allows connecting the ACs to imple-ment molecules. To be able to implement molecules with a rather high molecule-level parallelism (i.e., where multiple atoms execute in parallel, see Sect. 3.2), the communication infrastructure needs to provide sufficient parallel communi-cation channels.
3. Provide local storage to save the temporary computation results from atoms. This allows using temporary calculated results at a later point in time, which is especially important for molecules with a rather low molecule-level parallelism. In such a molecule, the atoms execute in a rather sequential way and thus the results are typically not demanded immediately but they are demanded at a later point in time.

All three points are important to provide modular SIs that support upgrading the SI implementations. In literature, general approaches exist that aim to provide a flexible infrastructure to allow interaction and dynamic communication between different reconfigurable modules. The Erlangen slot machine (ESM) [BMA+05, MTAB07] provides a flexible platform that provides multiple reconfigurable

regions with different communication mechanisms. However, the ESM cannot be
used to solve the specific problem of implementing modular SIs, because the provided intermodule communication does not fulfill the requirements: the ESM provides sufficient communication between adjacent modules, but nonadjacent
modules communicate either via a reconfigurable multiple bus (RMB) [ESS+96] or
(in rare cases) via the external crossbar. The external crossbar comes with the drawback of sending the data off-chip, which limits the achievable latency and bandwidth. The RMB instead is not suitable for rapidly changing communication
patterns (required for modular SIs as indicated for the SATD example in Fig. 3.4,
as a communication path first has to be established in a *wormhole* fashion before it
can be used efficiently. The DyNoC approach [BA05, BAM+05] proposes an interconnection network that supports a 2D temporal placement of reconfigurable modules and provides scalability. It would deliver the bandwidth, but the average
communication latency is high and thus not suitable to implement molecules efficiently. Ullmann uses a bus [UHGB04a] and thus provides low latency but the
design does not provide high simultaneous communications between multiple modules, as required.

Figure 5.4 provides a first overview of the atom infrastructure that is presented
in this monograph to implement modular SIs (details in Sect. 5.4.1). The ACs are
connected with each other in a linear chain using interconnect boxes. In addition to
the ACs, some nonreconfigurable modules are connected to the chain as well.
Among others, these nonreconfigurable parts are used to establish the data memory
access, as shown later. This linear chain of ACs and nonreconfigurable modules is
connected to the GPR file of the core pipeline. To be able to receive four 32-bit
inputs from the register file in parallel, the interconnect boxes need to provide a
sufficiently large communication channel. However, the interface of the atoms (and
thus the AC) is independent of the general input/output of the atom infrastructure.
In the proposed implementation, the atoms receive two 32-bit inputs and provide

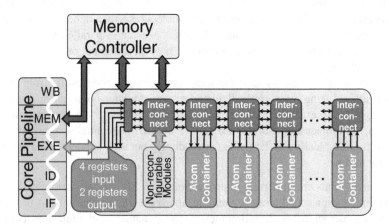

Fig. 5.4 Overview of the atom infrastructure and its connection to the core pipeline and the
memory controller

two 32-bit outputs.[8] In addition, each atom has a 6-bit input for control signals (e.g., reset or to choose some operation mode as DCT vs. inverse DCT) and an 8-bit output (notifying the system which atom is currently loaded into a particular AC). The interconnect boxes establishing the linear communication chain and provide access to the ACs. All ACs implement the same interface to assure that any atom can be reconfigured into any AC. However, the nonreconfigurable modules are not limited to any particular interface and thus they can implement specific connections to the atom infrastructure, depending on their communication requirements. In the concept of the RISPP architecture, only the ACs need to be implemented using a reconfigurable fabric. Although the prototype is realized using an FPGA (i.e., every part is implemented using a reconfigurable fabric), the core pipeline, memory controller, atom infrastructure, etc. will never be reconfigured during run time.

The interconnect boxes need to be reconfigured to determine which ACs should be connected with each other. In addition, the 6-bit control signal of each atom needs to be configured and also the nonreconfigurable modules demand some control signals. These configurations correspond to a coarse-grained reconfiguration of the atom infrastructure and it determines the calculations and operations (e.g., memory access) that shall be performed in a particular cycle. Therefore, it is required to change that configuration fast, i.e., from one cycle to another. Instead of using a fine-grained reconfigurable fabric to determine these configurations (because the reconfiguration time of these fabrics is rather long), a so-called very long control word (VLCW) is used to determine the configuration at a particular cycle. This means that the atom infrastructure corresponds to a coarse-grained reconfigurable design, and a VLCW corresponds to a particular configuration context that can be reconfigured in one cycle. An SI execution typically demands multiple cycles and in each cycle potentially different ACs need to communicate with each other, as shown in the SATD example in Fig. 3.4. Therefore, an SI execution demands multiple VLCWs, typically one per execution cycle. In the RISPP prototype, 1,024-bit VLCWs are used that are stored in a 64 KB large context memory (1,024 bits/VLCW × 512 VLCWs, implemented with 32 on-chip BlockRAMs). One 1,024-bit read port of the VLCW context memory is dedicated to the atom infrastructure to be able to load a complete VLCW per cycle. A 32-bit write port is connected to the run-time system to be able to change the content of the VLCW context memory. The execution of an SI in hardware[9] is determined by a start address that points to an entry in the context memory and the amount of subsequent VLCWs that shall be provided to the atom infrastructure (typically one per cycle). This coarse-grained reconfigurable atom infrastructure provides the flexibility to establish any communication pattern that the provided hardware supports (presented in Sect. 5.4.1). The execution of an SI using the cISA molecule or a mixed molecule (see Sect. 5.2) does not require any entries in the VLCW context

---

[8] Note that atoms do not necessarily use all inputs and outputs.

[9] Despite providing the input from and writing back the results to the general-purpose register file.

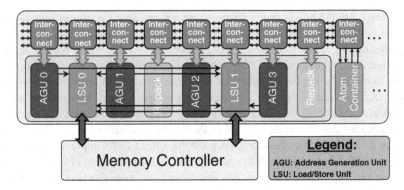

Fig. 5.5 Overview of the nonreconfigurable modules within the atom infrastructure

memory. However, the mixed molecules demand elementary SIs (using exactly one atom for one calculation) which require one VLCW entry per elementary SI.

Before investigating the architecture for the interconnect boxes and the ACs in Sect. 5.4.1, an overview of the nonreconfigurable modules that are connected to the atom infrastructure is given in Fig. 5.5. Three different types of nonreconfigurable modules are used:

1. *Load/store unit (LSU)*. Two LSUs are used to access two independent memory addresses in parallel (using the two independent memory ports as described in Sect. 5.3). Each LSU may load or store 128 bits at the same time. The address for the memory access can be obtained from the interconnect chain (e.g., as result of an atom computation or as input from the GPR file from the core pipeline) or from a dedicated address generation unit (AGU). The details of the LSU are described in more detail in Sect. 5.4.2.
2. *Address generation unit*. The AGUs are used to provide addresses to the LSUs. Each AGU can be initialized by values from the communication infrastructure, e.g., provided by the GPR file from the core pipeline. They can provide address sequences that describe certain patterns in the memory. For instance, a two-dimensional subarray of a larger two-dimensional array (e.g., a MacroBlock in a video frame) can be accessed. Altogether, four independent AGUs are provided. They can be used to provide two load streams and two store stream (or four load streams, etc.). Section 5.4.2 provides the details for the AGUs.
3. *Repack atom*. In addition to the LSUs and AGUs, also two instances of the Repack atom are provided as nonreconfigurable modules. They implement different byte rearrangement operations and the simulations suggest that these atoms are seldom replaced, as their functionalities are demanded by many SIs. Therefore, it is not necessarily beneficial to provide them as reconfigurable atoms. Instead, they are implemented as nonreconfigurable modules, which allows changing the interface. Four inputs and four outputs are provided (instead of two for reconfigurable atoms) to perform more operations in parallel or to perform more complex repacking operations that demand more inputs. Repacking

does not perform any calculation, but it rearranges data from an input source. For instance, the four lowest bytes of four different 32-bit words may be concatenated to form on output word. Such operations are demanded to prepare data for subsequent atom operations, i.e., they enable the calculations for the other atoms (e.g., a transformation). Even though no calculation is performed, executing such operations using the cISA corresponds to a rather long sequence of masking ("and" operation), shifting, and combining ("or" operation) instructions. In addition, the Repack atom has the smallest hardware requirements in comparison to other atoms, i.e., the amount of reconfigurable fabric within an AC would not be utilized efficiently when offering the repack functionality as reconfigurable atom. Providing the Repack atom statically increases the area efficiency correspondingly. In addition to the repacking operations, also an adder is included within the Repack atom, as some SIs demand accumulation of temporary results, which can be done using the nonreconfigurable adder.

5.4.1 Atom Containers and Bus Connectors

The atom infrastructure is partitioned into so-called atom containers (ACs) and bus connectors (BCs). The BCs correspond to the interconnect boxes in Fig. 5.4. Each AC is connected to a dedicated BC via so-called bus macros [Xil05b] to establish communication between the partially reconfigurable AC and the nonreconfigurable BC. A bus macro is a relatively[10] placed and routed IP core that provides a dedicated communication point. This is important for partially reconfigurable designs, as the static and the reconfigurable parts of the design are implemented after each other. When implementing the static part, only a free space for an AC is reserved and the communication interface is fixed (using the bus macros). Afterwards, multiple different atoms can be implemented for a particular AC, using the bus macros to communicate with the BC. The bus macros provide a communication interface to which the BS design can route its communication wires. Without bus macros, this routing point would depend on the atoms that can be reconfigured into an AC and in their specific implementation (not known during implementation of the static design).

For technical reasons, the reconfigurable regions have a rectangular shape. The smallest reconfigurable element in a Xilinx Virtex FPGA is a so-called frame [Xil05b], i.e., a frame is an individually addressable part of the configuration bits that determine the FPGA functionality. Multiple frames need to be reconfigured to reconfigure a configurable logic block (CLB). In Virtex-II devices, a frame covers

---

[10] This means that all components within the IP core are placed and routed relative to each other, but the IP core itself can be placed at different (not necessarily all) places on the FPGA without affecting the IP-core internal composition.

the whole height of an FPGA. Therefore, the smallest meaningful[11] reconfigurable region is a CLB column that spans the full FPGA height. However, it is possible to create reconfigurable modules that do not span the full FPGA height. Still the entire CLB column will be reconfigured although only a part of it will change its content. The other parts will obtain the very same configuration data that was loaded into them statically. Therefore, technically they are reconfigured (which also demands time) but practically they never change their content.[12,13] It is also possible to place multiple reconfigurable regions in the same column, using a read–modify–write strategy to perform the so-called 2D-partial reconfiguration [HSKB06, SKHB08]. Before a reconfiguration, the configuration bits of the corresponding frames are read, the parts of the configuration bits that contain the reconfigurable region are modified, and the frames are written back. This assures that the static part and all other reconfigurable parts within the frames maintain their previous configuration. The Virtex-4 (and later) devices provide frames that no longer span the full FPGA height, i.e., they naturally support multiple reconfigurable regions within the same CLB column.

Multiple AC–BC pairs are connected to collectively build the atom infrastructure. The partitioning into ACs and BCs makes sure that the BC functionality is available even during the reconfiguration of the AC. The BC has input and output connections to the adjacent BCs and to its AC. Figure 5.6 shows the internal composition of a BC in detail. Each BC contains two small local storages (one per atom output) with one write port, two read ports, and the capacity to store four temporary atom results. The inputs to the AC can be received from any of the BC inputs (four from each left and right adjacent BC) or from any port of the two local storages. Both AC inputs have the same possibilities from where the input shall come and both decide independently about their source using two independent multiplexers. A transparent latch is placed between the multiplexer and the AC input. This allows disconnecting the atom inside the AC from external inputs when it is not demanded and thus avoids unnecessary toggles and power consumption of the atoms. The AC output is written to the local storage, however, the input data to the ACs may also be written to the local storage directly (i.e., without passing through the atom). This allows copying data from a local storage to another one, which is required in case the provided size of the local storage is too small for a particular molecule execution. The output of the local storage can drive any output of the BC. The computing

---

[11] When only minor changes need to be reconfigured, then reconfiguring a single frame might be sufficient (e.g., to change a value in a LUT), however, reconfigurable modules typically cover CLB subarrays as smallest entity.

[12] The Xilinx Virtex series provides the feature of a so-called glitchless reconfiguration, i.e., if a configuration bit after its reconfiguration has the same value like before its reconfiguration, then it is guaranteed that this bit does not glitch in between (the Xilinx Spartan series does not provide this feature).

[13] Actually, the configuration is reset to the configuration of the static bitstream which might have changed since its initial configuration as a normal part of operation if, for instance, the configuration bits were used as memory using the so-called 'distributed RAM' IP core [Xil05a].

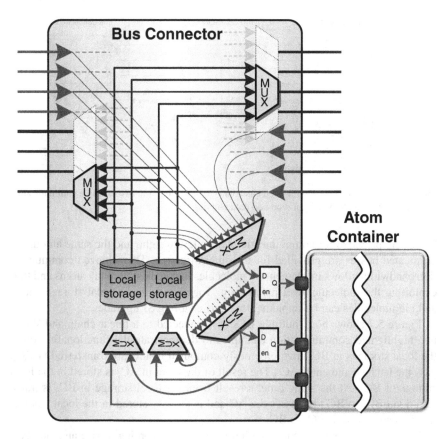

Fig. 5.6 Internal composition of a bus connector, showing the connection to its atom container and the neighboring bus connectors

results may reside in the local storage for multiple cycles. This is required to implement molecules that consist of a small amount of atoms. These molecules do not exploit parallelism exhaustively and therefore need to store intermediate results temporarily.

One of the main requirements of the communication infrastructure is to provide low communication latency while at the same time achieving high-communication bandwidth (e.g., the evaluated SATD molecule in Fig. 3.4 requires up to 256 bits per cycle, e.g., 8×32 bits in cycles 11, 14, and 15). Thus, segmented busses were implemented to connect the BCs in order to achieve a single-cycle communication (i.e., no registers on a path from local storage to AC) and to use the available bus lines efficiently [MSCL06]. Therefore, each bus can be used to establish multiple communications if the communication patterns do not intersect. For instance, in a linear chain of BCs (BC_1, BC_2, ...), a communication from BC_1 to BC_3 and another communication from BC_3 to BC_4 can be realized on the same segmented bus. To allow a parallel communication, unidirectional busses are used, i.e., multiple segmented

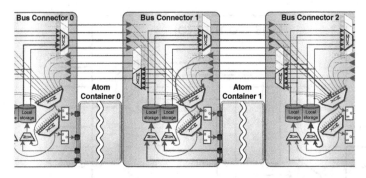

Fig. 5.7 Atom infrastructure with three bus connectors, highlighting two example communication patterns for typical atom computation and data copying

busses are provided for communications from left to right, and the same amount of segmented busses are provided for the other direction. The achieved communication bandwidth allows implementing molecules that consist of many atoms and thus exploiting the molecule-level parallelism. Together with the local storage, small and big molecules can be implemented, thus enabling SI upgrades.

Figure 5.7 shows how multiple BCs are connected to a linear chain and shows two highlighted communication examples to demonstrate the functionality. From the local storages in BC_2, two previously computed results are transferred to BC_1, pass the latches, and enter AC_1. The result of the atom in AC_1 is stored in the local storage of BC_1. At the same time, a result from the local storage in BC_0 is transferred via BC_1 to BC_2, bypasses the AC and is directly stored in the local storage, i.e., it is copied from BC_0 to BC_2.

Relevant parameters for the atom infrastructure – according to size and latency – are the number of buses and the bus granularity (e.g., 16-bit or 32-bit buses). For instance, if a 16-bit bus granularity is used, the Repack atom for SATD in Fig. 3.4 is no longer required for implementation. For this SI, the Repack atom is rearranging data on 16-bit level. When a 16-bit bus granularity would be used, this rearrangement could directly be implemented with the multiplexers that determine the atom input data (see Fig. 5.6). However, a 16-bit bus granularity comes at higher hardware cost and still the Repack atom is required for 8-bit data rearrangement for other SIs.

The atom infrastructure that is presented in this monograph is evaluated for different parameter values and presents detailed results of the implemented and tested FPGA implementation. The most relevant parameters are the amount of buses with their granularity and the quantity of BCs. A tool was developed that automatically generates the atom infrastructure (including the BCs and a constraint file for the bus macro placement) for different parameter settings to explore different design points.

Figure 5.8 shows the impact of the bus granularity on the hardware requirements (*y*-axis) for increasing amount of buses (*x*-axis). Thereby, *n* 32-bit buses are compared with $2n$ 16-bit busses to make sure that the compared bus systems provide the same bit width. As stated beforehand, offering a finer bus granularity may potentially

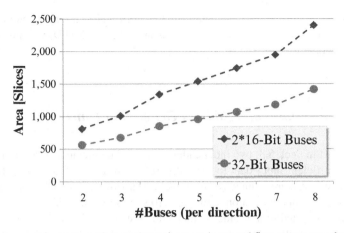

Fig. 5.8 Area requirements per bus connector for more buses and finer access granularity

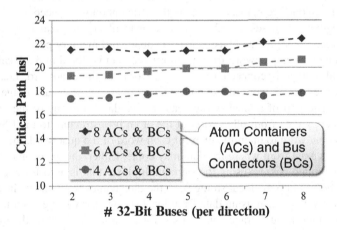

Fig. 5.9 Latency changes for increasing amount of buses and bus connectors

avoid the repacking of data. However, even for a moderate amount of four segmented buses, the hardware costs of 16-bit granularity are 1.58× bigger than that of 32-bit granularity (resulting in 492 additionally required slices per BC) and the gap increases further. Therefore, 32-bit granularity is selected for implementation.

Figure 5.9 analyzes the impact of different quantities of buses and BCs on the critical path. The amount of buses (x-axis) only affects the critical path (y-axis) slightly, but each additional BC in the bus length (different lines in the figure) increases it by approximately 1 ns. The critical path was analyzed for all shown measurements. In all cases, the critical path started reading data from the local storage of the rightmost BC, passing it to the leftmost BC (communication), through the leftmost atom (computation), and storing the results in the corresponding local storage. For this analysis, the atom with the longest atom-internal critical path (i.e., the Transform atom that is also used by SATD) was placed in all ACs to provoke this worst-case behavior.

5.4.2 Load/Store- and Address Generation Units

As explained in Sect. 5.3, two independent 128-bit data memory ports are provided to obtain sufficient input data to exploit parallelism. Figure 5.5 showed how two independent LSUs are connected to the atom infrastructure to access the memory and provide the data to the atoms. Figure 5.10 shows the internal composition of an LSU. It is connected to the same segmented busses as the bus connectors (BCs) of the atom infrastructure. However, in comparison to the BCs, the LSU contain four local storages (providing space for four different 32-bit words). The reason is that the LSU connects to a 128-bit (quadword) memory port. In order to be able to store a quadword in one cycle, local storage with sufficient bit width need to be provided. As the segmented busses are 32-bit wide, four local storages realize a good interface between the segmented busses and the quadword memory port.

When loading a quadword from memory, the data is stored in the local storage. Each local storage is connected to all four words of the quadword, i.e., it is not predetermined which word of a quadword is stored in which local storage. In addition, the write port of a local storage can receive data from any of the segmented busses to copy data from other parts of the atom infrastructure to the LSU. When writing a quadword to memory, then for each word of the quadword, the following possibilities exist from where the data may come. Either, the data comes from any of the segmented busses (using the same multiplexer that is used for copying data from the busses into the local storage), or the data comes from one dedicated local storage. However, when using data from local storage for writing, then it is predetermined which local storage can be used for a particular word of the quadword, i.e., not all local storages are connected to any word of the quadword at the memory write port. This was not needed, as the data that is written to the local storages has all flexibilities to be written to that local storage that corresponds to the later write position. Whenever a load or store access is performed, then the address may be obtained from the segmented busses or from the AGUs. Each AGU provides two addresses, which is important to utilize both LSUs as shown later. In addition to the shown signals in Fig. 5.10, also a byte select signal is send to the memory port. It chooses which of the 16 bytes in the quadword shall be accessed, which is especially important for write accesses.

The AGUs are used to calculate addresses in a certain pattern, i.e., they describe a so-called memory stream. For describing a memory stream, a base address is used in addition to the parameters stride, span, and skip, as described in Ciricescu et al. [CEL+03] and Lopez-Lagunas and Chai [LLC06]. Figure 5.11 shows an example of a memory stream and how it can be described using these parameters. The example shows a 2D array of data that contains a 2D subarray that shall be read by

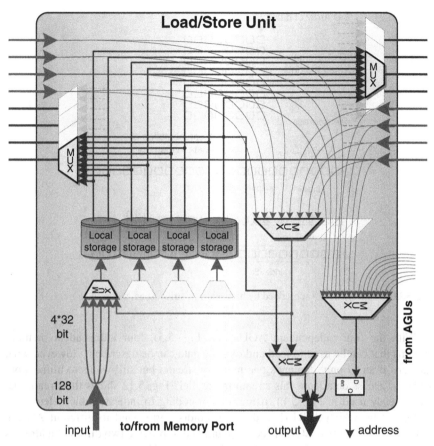

Fig. 5.10 Internal composition of a load/store unit, showing the connection to the memory port and the address generation units

the LSU. In memory, the 2D array is stored in a linear fashion, i.e., the rows of the array are stored sequentially after each other. The base address points to the first value of the subarray and the stride describes the increment to obtain the next element. In this example, the 2D subarray is dense, i.e., there are no gaps in between, and thus the second element is directly following the first one (i.e., the stride is one). The parameter for the span describes the size of a row, i.e., how many elements shall be accessed by adding the stride to reach the next element. After an entire row is read, the skip describes how the first element of the next row can be reached from the last element of the current row. From there, the span determines how many elements shall be read before the skip needs to be applied again. These parameters also allow traversing the 2D subarray in a vertical manner, i.e., reading it columnwise. Therefore, the stride describes how to come from the ith element of a row to the ith element of the next row. The span describes the number of rows and the skip resets the address to the next element of the next column.

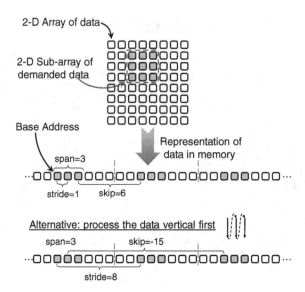

Fig. 5.11 Memory streams, described by base address, stride, span, and skip

With the four independent AGUs (see Fig. 5.5), four independent memory streams that can be used to load and/or store data can be described. However, what happens if an SI only demands one memory stream, but still wants to utilize both LSUs to access data from this stream in parallel? Fig. 5.12 shows the stream that was already used in Fig. 5.11, using a color-coding to indicate which element is addressed by which AGU (i.e., different colors correspond to different AGUs). Obviously at least two AGUs need to be utilized to provide two different addresses in parallel. In the first attempt in Fig. 5.12a, two AGUs are used and the elements are colored alternating red and green to indicate that the addresses of two successive elements are available at the same time. However, the resulting pattern for the red- and green-colored elements can no longer be described using the stride, span, and skip parameters, respectively. The problem is that an odd number of elements belong together in a group and thus two different kinds of groups are obtained, i.e., one with two red elements and one with two green elements. Therefore, Fig. 5.12b shows the second attempt that uses all four AGUs to describe one memory stream. Two AGUs are used to describe the group with the two red elements, and the other two AGUs are used to describe the other group (now with two blue elements). Figure 5.12c shows the resulting AGU to LSU mappings over time, i.e., at which time, which AGU provides an address to which LSU.

If an SI needs one input stream and one output stream and want to use both LSUs to read the input data, then all four AGUs are busy to create the addresses for the input stream. Therefore, the output stream cannot be written until all input data is read. This situation requires a rather large amount of temporary storage to buffer the input and/or output data until the input stream is processed entirely. Depending on the available amount of local storage, it might even be impossible to execute the

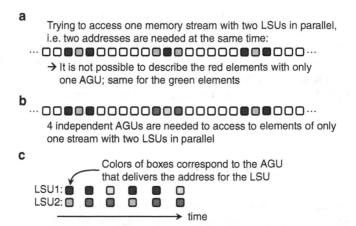

Fig. 5.12 Problems, if one memory stream shall be accessed with both LSUs in parallel

SI using both LSUs. For instance, if 256 quadwords shall be read and for each 16 quadwords one result is written back to memory, then the local storage would need to buffer at least 256×4 words = 4 KB (noticeably larger than the local storage that is provided in the atom infrastructure). And even if sufficient buffer would be available, still the parallelism is limited, as the store operations may not start until all load operations are completed (even though results that shall be stored are already available). The reason is that at least one AGU needs to be reinitialized to provide the addresses for the store stream. In order to provide a more efficient solution for this situation, the work that is presented in this monograph proposes to extend the AGUs to provide two addresses in parallel. This means that each AGU calculates the next two addresses of its stream in two subsequent cycles and offers both in two address output registers. If only one address is requested, then only one new address is calculated. If both LSUs request an address, then it demands two cycles until the AGU has calculated the next two addresses. As the LSU access to memory demands at least[14] two cycles (one to request the data and one to acknowledge its reception and store it in the local storage), the AGU finishes calculating the next two addresses in time, i.e., the LSU can immediately start the next access.

The AGUs are also connected to the segmented busses of the atom infrastructure (see Fig. 5.5). However, the AGU never writes to them but only reads their values to initialize the base address, stride, span, and skip. As four AGUs are provided but only four registers may be read from the GPR file from the core pipeline, it would only be possible to initialize four independent base addresses if all four AGUs are used. Therefore, dedicated extra bits are reserved in the VLCW to determine the configuration of the atom infrastructure (see Sect. 5.4). The VLCW contains 11 bits

---

[14] In case of an unaligned access it might take longer, same if the LSU accesses the main memory instead of the on-chip scratchpad memory.

per AGU that can be used – in addition to the values that come from the segmented busses – to initialize the parameters, stride, span, and skip. Six additional bits per AGU are used to determine which combination of the above-mentioned 11 bits and the input from the segmented busses shall determine the stride, span, and skip. However, these bits in the VLCW are constant, i.e., they are determined when the SI is designed at compile time. The input from the segmented busses typically comes from the GPR and thus it can be changed at run time.

5.4.3 Summary of the Atom Infrastructure

The atom infrastructure connects the core pipeline with the run-time reconfigurable atoms and provides a communication and computation infrastructure that allows implementing different molecules. This monograph presents segmented busses for a high-bandwidth communication and local storage to save intermediate results. This allows implementing molecules with rather small molecule-level parallelism (demanding more local storage) and with rather high molecule-level parallelism (demanding more communication). Two LSUs and four AGUs are connected to the segmented busses. They allow accessing the main memory and an on-chip scratchpad with two independent 128-bit ports and provide a high flexibility in describing the memory streams that shall be accessed. In addition, data repacking functionality is provided as a nonreconfigurable unit within the atom infrastructure. A design-space exploration suggested that offering the segmented busses at the access granularity of 16 bits would result in a noticeable hardware overhead. Therefore, the functionality of data repacking is required. Simulations showed that corresponding Repack atoms were seldom replaced and thus they can be provided in nonreconfigurable logic. The atom containers (ACs) in the atom infrastructure provide a fine-grained run-time reconfigurable fabric that can be reconfigured to contain an atom. A VLCW configured the atom framework for each cycle of an SI execution. This corresponds to a coarse-grained reconfiguration of the atom framework that determines how the fine-grained reconfigurable ACs shall be connected to implement the functionality of a molecule.

5.5 RISPP Prototype Implementation and Results

The RISPP prototype is implemented on a Xilinx Virtex-4 LX 160 FPGA, using a board from Silica/Avnet (see Appendix B for details on the FPGA board and the developed PCB). The early access partial reconfiguration (EAPR, [Xil08a]) tool flow is used with ISE 9.1.02i_PR10 and PlanAhead 10.1. The Leon2 processor [Aer] was used as core pipeline. The online monitoring, forecast fine-tuning (see Sect. 4.3), the special instruction (SI) and helper instruction opcodes/formats (see Sect. 5.1), the atom infrastructure (see Sect. 5.4), and a special IP core for performing partial reconfiguration via the internal configuration access port (ICAP, [Xil09c]) are implemented in hardware. The algorithms for molecule selection (see Sect. 4.4), reconfiguration-sequence scheduling (see Sect. 4.5), and atom replacement (see Sect. 4.6) are implemented in software, using a MicroBlaze [Xil09b] soft-core processor that is connected to the system. As already indicated in Fig. 4.4, these algorithms only need to be triggered by the core pipeline and then can execute in parallel to it without frequent synchronization. Implementing these algorithms in software provides the advantage that it is relatively easy (in comparison to a hardware implementation) to modify them and to obtain debugging and status information. This allows examining different alternatives in a rather short time.

Figure 5.13 provides an overview of the part of the run-time system that is controlled by the MicroBlaze. In addition to the typical periphery modules for external RAM, debugging, and peripherals, it was extended to provide RISPP-specific interfaces. The connection to the core pipeline allows that the MicroBlaze can access the forecast values whenever a new forecast block executes, i.e., the execution of the algorithms that run on the MicroBlaze can be triggered (more details on their execution time is given below). In addition, this interface allows activating new molecules, i.e., the decode stage of the core pipeline contains a small lookup table that contains the information which molecule shall be used to implement an SI, and the MicroBlaze can write this information, e.g., after an atom finished reconfiguration or was replaced. Before a new molecule can be activated, the corresponding

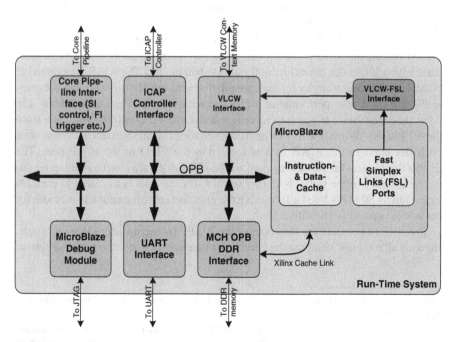

Fig. 5.13 Overview of the MicroBlaze system that implements the algorithms of the run-time system and controls the reconfigurations

VLCWs for its implementation need to be written to the dedicated VLCW context memory. To achieve a low latency and high bandwidth write access from the MicroBlaze to the VLCW context memory, the fast simplex link (FSL, [Xil09a]) interface is used to send data to the context memory.

To be able to reconfigure an atom, a dedicated IP core was developed that is connected to the MicroBlaze and that reads the partial bitstreams from an external EEPROM from Samsung [Sam05], buffers a part of the data in a FIFO, and streams the data to the ICAP port for reconfiguration. The KFG5616 OneNAND that was used in this work provides 32 MB of data[15] that is partitioned into 512 blocks, comprising 64 pages with 1 KB per page. Although the partial bitstreams for the atoms are smaller than 64 KB, a 64 KB block is dedicated to a particular partial bitstream (altogether allowing to store 512 different partial bitstreams in the 512 blocks) to simplify data management. A PCB was designed to connect the OneNAND EEPROM (in addition to further peripherals) to the RISPP FPGA prototype (see Appendix B). The IP core that was developed in the work presented in this monograph controls the OneNAND EEPROM and is connected to the MicroBlaze. The MicroBlaze can start a reconfiguration by providing the information, which atom shall be loaded into which AC (this corresponds to a particular address in the EEPROM) and the information about the length of the bitstream. The IP core performs the reconfiguration in parallel to the MicroBlaze execution and returns a checksum of the reconfigured bitstream. The EEPROM provides access to a double-buffer SRAM interface, i.e., during reading the content of a buffer from the EEPROM (temporarily stored in an on-FPGA FIFO), the EEPROM can be instructed to automatically load the other SRAM buffer with data from the EEPROM-internal NAND-array.[16] However, after reading the content from one SRAM buffer, the process has to wait until the other SRAM buffer is filled with data, which limits the overall reconfiguration bandwidth. The ICAP is operated at 50 MHz (same frequency as for the core pipeline and the MicroBlaze), but the average[17] EEPROM read performance limits the achieved reconfiguration bandwidth. After the MicroBlaze triggered a reconfiguration, the FIFO is filled with data from the EEPROM. When the FIFO contains sufficient data to assure a continuous 50 MB/s burst to the ICAP, data is sent from the FIFO to the ICAP port. The EEPROM delivers the remaining parts to the FIFO in parallel to the running reconfiguration. Due to the initial buffering (until sufficient data is available to perform a continuous 50 MB/s burst afterwards), the effective reconfiguration bandwidth for the whole process is 36 MB/s.

Figure 5.14 presents the floorplan of the RISPP prototype after place and route, showing all modules including the Leon2 core pipeline (bottom), the run-time system

---

[15] In addition, a spare memory area is available that is used to mask EEPROM-internal errors of memory cells.

[16] A special flash memory type (which is a special EEPROM type) that provides faster access time and a smaller footprint in comparison to a NOR-array, but does not offer a random access on the data.

[17] Considering the average time to read an SRAM buffer and wait until the other buffer is filled.

(MicroBlaze on left side), and the atom infrastructure (atom containers as empty boxes; bus connectors are between them). Most of the modules use a contiguous region. However, in some cases, they are partitioned. For instance, most parts of the Leon2 are placed in lower edge of the FPGA, but some parts are placed in between the ICAP controller and the I²C periphery. In early attempts, the design was floor-planed manually, by defining rectangular boxes for the modules. However, these attempts typically did not fulfill the timing constraints (50 MHz for core pipeline and MicroBlaze, and individual constraints for the paths of the atom infrastructure). After allowing an automatic placement of the nonreconfigurable modules, these small regions that are disconnected from the majority of the other components of the same module (like in the case of Leon2) allow that the timing constraints are met.

The atom containers (ACs) are shown as empty rectangular regions in Fig. 5.14. Their height corresponds to one Virtex-4 frame (i.e., 16 CLBs) and they are nine CLBs wide, providing 576 slices per AC. They appear to be empty, because only the placed logic of the static design is shown. However, even in the static design, the ACs are not empty, as the static design may use routing resources (not shown in the figure) of the AC to establish communication between different parts of the static design. Due to these already used routing resources, the bitstreams of the atoms that are placed and routed for different ACs have different sizes, as shown later. The bus macros – connecting the ACs with the atom infrastructure – are also not directly visible in the floorplan. However, they become apparent as small white gaps in the floorplan, directly adjacent to the ACs. In the shown floorplan, the ACs are partitioned into two columns, where each column is interrupted between the second and third AC. In early attempts, the two AC columns were connected and the bus connectors (BCs) were placed in between. However, as the Virtex-4 FPGA provides the BlockRAMs only on the left and right side of the device, rather many routing resources are required to provide the VLCWs from the BlockRAMs to the BCs. The connected AC columns acted like a barrier to between the BCs and the BlockRAMs and thus the tools had problems in routing it (either not routable or timing constraints failed). Therefore, the AC columns are disconnected such that the BC can be placed closer to the VLCW context memory to assure routability and timing.

Table 5.6 shows the hardware implementation results for the static design. The hardware requirements of the run-time system are moderate (dominated by the MicroBlaze that is used for development and debugging purpose). However, the atom infrastructure implies a noticeable overhead in comparison to the core pipeline. At the same time, the RISPP concept provides a noticeable performance improve-ment in comparison to state-of-the-art reconfigurable and nonreconfigurable pro-cessors, as is evaluated in Chap. 6. The multiplexers that are used to connect the modules with segmented busses (i.e., the BCs and the multiplexers in the static Repack atom and the LSU) dominate the size of the atom infrastructure. Here, it has to be recalled that the RISPP architecture is envisioned to be an ASIC design with an embedded FPGA (like presented and analyzed in Neumann et al. [NvSBN08] and Sydow et al. [SNBN06]) and for prototyping purpose all parts are implemented on an FPGA. Ehliar and Liu analyzed the hardware implementation

Periphery IP-
Core for I²C
(touch-screen
LCD)

ICAP
Controller:
external
EEPROM→
FIFO →ICAP

Bus Macros

MicroBlaze (for
Selection,
Scheduling and
Replacement)
and Peripherals

Load/Store
Unit 1

Load/Store
Unit 0

Address
Generation
Units

Leon2 core
Instruction Set
Architecture

Atom
Containers

Bus Connectors
and static
Repack Atoms;
parts of the
Atom
Infrastructure

Periphery IP-
Core for Video-
In and Video-
Out. Additi-
onally
providing video
buffers and
memory-
mapped
interface to
access the
buffers

Memory
Controller

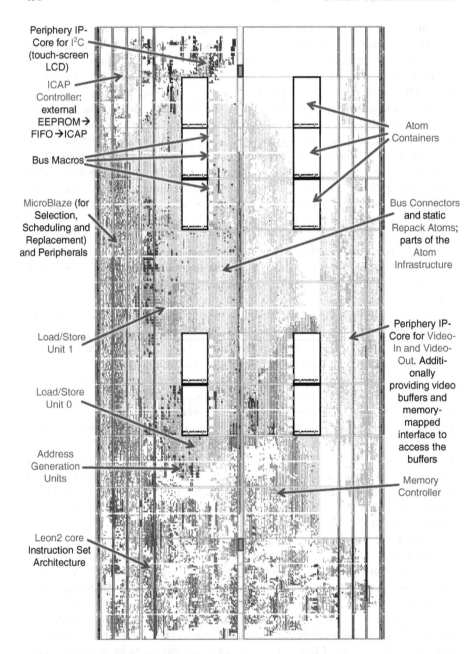

Fig. 5.14 Floorplan of the RISPP prototype implementation, showing the placement of the different components on the Xilinx Virtex-4 LX 160 FPGA

Table 5.6 Hardware implementation results for the RISPP prototype; all FPGA utilization numbers are relative to the used Xilinx Virtex-4 LX 160 FPGA

Module	Slices/FPGA utilization	LUTs/FPGA utilization	FFs/FPGA utilization	BlockRAMs/FPGA utilization	DSP blocks/FPGA utilization	I/O pins
Leon2	4,366/6%	7,157/5%	2,160/2%	13/5%	–	SD-RAM (57), UART (2)
Repack atom[a]	2,004/2%	3,284/2%	128/<1%	–	–	–
Bus connector[a]	1,040/2%	1,704/1%	64/<1%	–	–	–
Address generation unit[a]	400/<1%	655/<1%	124/<1%	–	–	–
Load/store unit[a]	1,912/2%	3,132/2%	65/<1%	–	–	–
VLCW context memory	56/<1%	2/<1%	90/<1%	32/11%	–	–
Total atom infrastructure	19,880/29%	32,498/24%	1,740/1%	32/11%	–	–
MicroBlaze	4,384/6%	7,184/5%	4,734/4%	10/3%	3/3%	DDR-RAM (107), UART (2)
Monitoring, fine-tuning, and FI to MB interface	640/1%	1,053/<1%	922/<1%	10/3%	2/2%	–
ICAP controller	540/1%	885/1%	608/<1%	16/6%	–	EEPROM (40)
Total run-time system	5,564/8%	9,122/7%	6,264/5%	36/13%	5/5%	DDR-RAM (107), UART (2), and EEPROM (40)
Memory controller	6,308/9%	9,723/8%	2,211/2%	–	–	–
Scratchpad	–	–	–	16/6%	–	–
I²C periphery	864/1%	1,416/1%	860/<1%	3/1%	–	Touch screen LCD (15)
Video module periphery	4,372/6%	7,164/5%	680/<1%	102/35%	4/4%	SRAM (67), video extension board (41)

cost for multiplexers, comparing FPGA targets with an ASIC target [EL09]. They compared the implementation cost for a 32-bit 16:1 multiplexer (MUX) with a 32-bit adder and presented the area requirements of the MUX relative to the adder requirements of the same technology. For Xilinx FPGAs, the MUX is between 5.0× (Virtex-5) and 8.0× (Spartan-3[18]) larger than the adder. However, for 130 nm ASICs, the MUX only demands between 0.48× and 0.57× of the adder area, i.e., the multiplexers of the atom infrastructure will benefit significantly from an ASIC implementation, diminishing the reported area overhead. However, as no embedded FPGA design is available, in the scope of the presented work, no synthesis and implementation results could be obtained to compare an ASIC implementation of the RISPP architecture with an ASIC implementation of the Leon.

The ACs – shown in Fig. 5.14 – can be reconfigured at run time to contain atoms. Table 5.7 shows the implementation results of the atoms that are implemented for the H.264 video encoder example that are used for benchmarking and comparisons. The largest atoms are the Clip3 atom (as it is extended to support various different data clipping operations) and the PointFilter (as it demands multiple 8-bit multipliers that are implemented with slices). The smallest atoms are QSub and CollapseAdd that perform rather simple addition and subtraction operations. The utilization of an AC is rather moderate (Clip3 uses 56% of the available slices), thus it would be possible to reduce the size of an AC. However, the design shall not be optimized toward a specific application and its requirements,

Table 5.7 Atom implementation results

Atom	Slices/AC utilization[a]	LUTs/AC utilization[a]	Critical path[b] [ns]	Bitstream size[c] [byte]	Reconfiguration time[d] [ms]
Clip3	252/56%	413/46%	9.8	30,335–33,370	0.91
CollapseAdd	36/8%	58/6%	7.4	18,940–25,709	0.70
LF_4	144/32%	236/26%	11.6	26,243–29,447	0.80
Cond	82/18%	132/15%	8.1	21,932–28,860	0.78
PointFilter	184/41%	300/33%	15.1	27,658–31,798	0.86
QSub	20/4%	32/4%	3.2	16,340–25,724	0.70
SADrow	104/25%	185/20%	13.0	24,337–29,057	0.79
SAV	58/13%	93/10%	8.4	22,912–28,928	0.78
Transform	124/30%	217/24%	7.5	25,454–30,242	0.82

[a]All AC utilizations are relative to the 9 × 16 CLBs large ACs that provide 576 slices and 1,152 LUTs
[b]The atom-internal critical path is constrained for each atom individually, however, independent from any particular AC
[c]The bitstream sizes are different, depending on the AC for which they are created
[d]For the largest bitstream size of that AC, using 36 MB/s reconfiguration bandwidth

[18] No comparisons with Virtex-4 are presented, but the Spartan-3 provides the same LUT structure than the Virtex-4 (i.e., four-input LUTs), whereas the Virtex-5 uses a redesigned structure (six-input LUTs); thus the Spartan-3 results are considered to be more significant for the presented Virtex-4 results.

instead it should be flexible to support other SIs and atoms after the static design implemented successfully. In Table 5.7, it is noticeable that the bitstream sizes vary for a given atom that is implemented for different ACs. These differences reflect the utilization of the ACs for nonreconfigurable routing, i.e., communications between two nonreconfigurable parts may use routing resources inside an AC (the EAPR tool flow assures that the communication between the nonreconfigurable parts is not affected by run-time reconfiguration of the AC [LBM+06]).

Except QSub (demanding 3.2 ns), the critical path of different atoms ranges from 7.4 to 15.1 ns (see Table 5.7). As the core pipeline is constrained to operate at 50 MHz (20 ns) in the RISPP prototype, it becomes obvious that not all operations in the atom infrastructure may be single-cycle operations. A typical operation was shown in Fig. 5.7, i.e., reading data from the local storage, sending that data over the segmented busses to the target BC, performing the computation within the AC, and storing the result in the local storage. Actually, the critical path even starts earlier, as the VLCW that determines which address from the local storage shall be read, needs to be provided from the VLCW context memory to the corresponding BC. Table 5.8 shows the individual parts of the overall critical path of the atom infrastructure in the sequence in which the parts are accessed. All these parts are constrained by individual timing constraints, which is necessary in partially recon- figurable designs if the critical path passes a reconfigurable region without registers in the bus macros. During implementation of the static design, the tools do not know which atoms might be loaded into the reconfigurable parts during run time and thus they cannot consider the actual critical path. Actually, the tools are not even aware that the AC input is connected to the AC output, as the ACs are just handled as a black box during implementation.

The latencies in Table 5.8 correspond to the individual timing constraints, i.e., the reported minimum and maximum latencies do not correspond to the results of a particular implementation run, but they are guaranteed for any implementation run (unless a timing error is reported). Step 1 is constrained to 5 ns for any BC that is driven by the VLCW. Step 2 depends on the distance of the two BCs that com- municate with each other, 12 ns for directly neighbored BCs and 22.5 for the most distinct BCs. In general, it requires approximately 1 ns to send the data from one BC to the neighbor BC. Step 3 corresponds to the atom computation (see Table 5.7), and Step 4 saves the computed results into the local storage (3.0 ns for most BC, but for some BCs the constraint had to be relaxed to 3.3 ns). The shortest path

Table 5.8 Individually constraint parts of the critical path in the atom infrastructure and their processing sequence

Step	Source	Destination	Minimum latency [ns]	Maximum latency [ns]
1	VLCW context memory	Bus connector	5.0	5.0
2	BC-internal local storage	AC input at target BC	12.0	22.5
3	AC input	AC output	3.2	15.1
4	AC output	Local storage	3.0	3.3
Sum	–	–	23.2	45.9

demands 23.2 ns and thus is longer than the critical path of the core pipeline. This means, that all operations in the atom infrastructure are multicycle operations. The longest path even demands 45.9 ns, however, nearly all combinations of operations require two cycles. Multicycle operations are realized by applying a particular VLCW for two (in rare cases three) cycles, i.e., the atom infrastructure keeps a certain VLCW configurations to assure that all operations are completed.

In addition to the hardware area- and timing results, the algorithm execution time of the run-time system is also benchmarked executing on the MicroBlaze. All algorithms described in Chap. 4 are implemented on the MicroBlaze, except the online monitoring and forecast fine-tuning (see Sect. 4.3) that are implemented in hardware. As visible in Fig. 4.4, the algorithms on the MicroBlaze (yellow box in that figure) are triggered by a forecast instruction. Afterwards, the molecule selection, reconfiguration-sequence scheduling, and atom replacement execute sequentially. However, in order to start the first atom reconfiguration, not all atom reconfigurations need to be determined in advance. For instance, the molecule selection determines the first molecule after one iteration of its outer loop and reconfiguring atoms for that molecule can already be started even though the other molecules are not selected yet. Actually, the reconfiguration-sequence scheduling might prefer loading atoms for another molecule first, however, waiting until the molecule selection determined all molecules before starting the scheduling, replacement, and finally the first reconfiguration is not necessarily beneficial.

Therefore, the algorithms for selection and scheduling are partitioned into two runs: the *first run* determines one molecule and schedules the atoms for that molecule. Afterwards, the replacement determines the replacement candidate and the reconfiguration can be started. The algorithm for replacement determines one replacement decision at a time, i.e., there is no need to partition it as well. After the first reconfigurations are started, the *second run* is started that determines all further molecule selections and schedules the additionally demanded atoms of all selected molecules. The algorithm execution time for the second run is hidden (to some degree) by the reconfigurations that are started by the first run. In the best case, only the algorithm execution time before starting the first reconfiguration appears as overhead of the run-time system, i.e., delaying the first reconfiguration. During the reconfiguration, the run-time system checks for new forecasts and aborts the running selection, etc. in case a new forecast appears.

In Fig. 5.15, a detailed algorithm execution time analysis is presented that was measured on the RISPP prototype (using performance counters that were added to the MicroBlaze) executing the three computational blocks, motion estimation, encoding engine, and in-loop de-blocking filter, of an H.264 video encoder (see Sect. 3.3). The x-axis represents different points in time by showing each execution of the outer loop of the run-time system as one bar. Please note that this does not correspond to a linear time axis, as these outer loop iterations have different execution times. One iteration of that outer loop typically determines one reconfiguration decision. For instance, the first run of the selection, scheduling, and replacement (as described above) corresponds to one iteration of the outer loop. If the molecule that was selected in the first run demands multiple reconfigurations, then performing

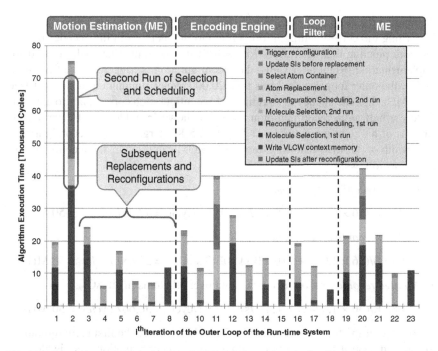

Fig. 5.15 Execution time analysis of the RISPP run-time system's algorithms that execute on the MicroBlaze part of the RISPP prototype

these reconfigurations (including a replacement decision beforehand) corresponds to the next iterations of the outer loop. After all these atoms are reconfigured, the second run of selection and scheduling is executed and the subsequent reconfigurations correspond to the following iterations of the outer loop. However, in case the first run selects a molecule that does not require any reconfiguration (i.e., all demanded atoms are already available), then this iteration of the outer loop does not trigger any reconfiguration and the next iteration directly corresponds to the second run of the selection and scheduling.

For each iteration of the outer loop, Fig. 5.15 shows the execution time of the different algorithms of the run-time system (that execute during that iteration) as stacked bars on the *y*-axis. The sequence of bars – from bottom to top – directly corresponds to their execution sequence in the iteration of the outer loop. The first iteration of the outer loop starts (from bottom up) by executing the first run of selection and scheduling and afterwards performs one replacement decision, selects an AC (hardly visible in the stacked bars), updates the elementary SIs, and triggers the first reconfiguration. The replacement determines which atom type shall be replaced, but in case multiple instances of the atom type that shall be replaced are available, it does not determine which instance (i.e., in which AC) shall be replaced. "Select AC" simply iterates over all ACs and chooses the first one that contains the corresponding type. The function "update SI before replacement" assures that all molecules demanding the atom in the AC that is going to be replaced are no longer executed, by writing the corresponding information into the memory in the decode stage of the core pipeline that determines for each SI the molecule that shall be used for implementation. Right afterwards, the reconfiguration is triggered. The first iteration of the outer loop required less than 20,000 cycles running on the MicroBlaze at 50 MHz, i.e., the latency to start the first reconfiguration corresponds to approximately a half atom reconfiguration time. As shown in Table 5.7, the partial bitstreams of an atom are between 18.5 and 32.5 KB large. For the reconfiguration bandwidth of 36 MB/s that the external EEPROM can provide, the reconfiguration time is as follows:

- *Longest reconfiguration time*:

  33,370 B/(36 × 1,024 × 1,024 B/s)/20E-9 s/cycle = *44,201 cycles* at 50 MHz

- *Shortest reconfiguration time*:

  18,940 B/(36 × 1,024 × 1,024 B/s)/20E-9 s/cycle = *25,087 cycles* at 50 MHz

- *Average reconfiguration time* (averaged over all atoms in all ACs):

  27,682 B/(36 × 1,024 × 1,024 B/s)/20E-9 s/cycle = *36,667 cycles* at 50 MHz

Right after the first loop iteration in Fig. 5.15 triggered the first reconfiguration, the second loop iteration starts in parallel to the reconfiguration. As visible, that loop iteration represents the peak algorithm execution time of all shown loop iterations that cannot be hidden by the atom reconfiguration time. The dominating parts correspond to writing the VLCW context memory and performing the second run of selection and scheduling. Writing the VLCW context memory is required to activate the molecules that become executable due to the atom that was reconfigured. Figure 5.15 shows the start of the application execution and thus no usable atom is available when the first execution of the motion estimation starts. Therefore, the first few atoms that are reconfigured lead to molecules with a rather slow execution time (their performance is upgraded when more atoms finish reconfiguration). For rather slow molecules, many entries need to be written to the VLCW memory, as one 1,024-bit VLCW entry is demanded per cycle of the molecule execution time. For the second execution of the motion estimation (writing the VLCW context memory starts in iteration 20), some atoms are already available and thus faster molecules are available immediately. Therefore, the time for writing the VLCW memory is significantly shorter (less entries need to be written). For the same reason, the execution time of the scheduling in iteration 20 is shorter, as the algorithm does not need to calculate a schedule that starts at an empty system (i.e., no atoms are available), but – as some usable atoms are available – less reconfigurations need to be scheduled.

The subsequent outer loop iterations of the run-time system (3–8) correspond to the reconfigurations that were planed by the second run (iteration 2). It is noticeable that none of these loop iterations demands more than 24,146 cycles, i.e., all of them are hidden by the reconfiguration, even if the smallest partial bitstream is reconfigured

(25,087 cycles). For the subsequent loop iteration, the first loop iteration of a new forecast (i.e., a new computational block at iterations 9, 16, and 19) is not hidden by concept (they trigger the first reconfiguration for that computational block). In addition, the loop iterations 11, 12, and 20 are not hidden if the smallest partial bitstream is reconfigured. However, in case the largest partial bitstream is reconfigured they are all hidden as well. This shows that the algorithm execution time of the run-time system does not affect the overall reconfiguration time significantly. Besides the start of the application (where no atoms are available), the execution time overhead of the run-time system corresponds to approximately half of the reconfiguration time of an atom. This overhead appears for the "first run" outer loop iteration of a new computational block and delays the start of the first reconfiguration.

5.6 Summary of the RISPP Architecture Details

Implementing the RISPP architecture started with an existing core pipeline (here the Leon2, but the concept is not limited to a particular core pipeline) and extended the instruction format to support special instructions (SIs) and helper instructions (HIs, accelerating parts of the RISPP concept and providing access to RISPP parameters). If the hardware for an SI is not reconfigured when the SI shall execute, then a trap handler executes the SI using the cISA. The cISA execution of SIs is one example that is accelerated by HIs. The register file is extended and a special memory access to an on-chip scratchpad memory is provided to obtain sufficient input data to exploit the molecule-level parallelism of SIs. To execute SIs in hardware, a special atom infrastructure was developed in the presented work to provide a communication and computation infrastructure that supports molecule upgrades, i.e., it supports molecules with rather few and with rather large molecule-level parallelism. The atom infrastructure contains two LSUs that have access to two independent 128-bit memory ports of on-chip scratchpad and it contains four AGUs that can describe four independent memory streams where each stream may be accessed in parallel by both LSUs. In addition, the atom infrastructure contains atom containers (ACs) that can be reconfigured to load atoms and that are connected to bus connectors (BCs). These BCs are connected with each other using four independent segmented busses (per direction) in a linear chain to establish communication between the ACs. In the scope of the work presented in this monograph, the entire RISPP architecture was implemented as a prototype (using the FPGA board shown in Appendix B), whereas the algorithms of the run-time system are implemented using a MicroBlaze soft-core processor. The hardware implementation results, area requirements of the individual components, critical paths of the atom infrastructure, algorithm execution time, and overhead of the run-time system are analyzed in detail. Altogether, the prototype implementation of the RISPP architecture demonstrates that the concept of modular SI is feasible and can be realized in practice.

Chapter 6
Benchmarks and Comparisons

In this chapter, the RISPP approach is benchmarked. The benchmark results for the individual parts of the run-time system and prototyping results for the RISPP architecture are already presented in Chap. 4 and in Sect. 5.5, respectively. The first section provides benchmarks for the entire RISPP approach for different architectural parameters. In addition, the RISPP approach is compared with state-of-the-art approaches. There are two different types of state-of-the-art approaches that need to be considered:

1. Nonreconfigurable application-specific processors (ASIPs, comparison in Sect. 6.3)
2. Reconfigurable processors with monolithic SIs (comparison in Sect. 6.4)

Both approaches use special instructions (SIs) and provide accelerators for their execution. In addition to these comparisons, this monograph provides a comparison with a general-purpose processor (GPP). A GPP is not considered as state-of-the-art approach in comparison to ASIPs and reconfigurable processors (that use the same core pipeline as the GPP and extend it with accelerators). However, related work often provides a comparison only with GPP and therefore this monograph presents these numbers to allow for later comparison.

For benchmarking, an in-depth analysis of an H.264 video encoder is performed as a detailed case study for different parameters instead of summarizing the results of multiple applications from MiBench [GRE+01] or MediaBench [LPMS97]. Note that many of the applications in these suites demand only one computational kernel [e.g., performing either discrete cosine transformation (DCT) or sum of absolute differences (SAD), or variable length coding (VLC)]. Many of these kernels are components of the H.264 encoder (H.264 demands further kernels like the Hadamard transformation that is not present in MiBench or MediaBench) and all of them are needed to be accelerated to expedite the entire video encoding process. However, when targeting applications that comprise only one kernel, then the concept of nonreconfigurable ASIPs typically provides the best performance and performance per area, as it can be tightly optimized for a specific kernel. Reconfigurable processors – as well as RISPP – mainly introduce flexibility in such a single-kernel scenario, i.e., they are also able to accelerate further application (e.g., in a multitasking environment), whereas the ASIP is dedicated to a single application.

L. Bauer and J. Henkel, *Run-time Adaptation for Reconfigurable Embedded Processors*,   165
DOI 10.1007/978-1-4419-7412-9_6, © Springer Science+Business Media, LLC 2011

However, for a rather large and complex application as the H.264 video encoder, the concept of reconfigurable processors may perform better than the ASIP because the hardware can be reconfigured for the individual kernels of the application during run time. This is the reason, why this monograph focuses on a detailed analysis of that challenging application, i.e., instead of benchmarking scenarios where reasonable solutions already exist in industry (e.g., [ARC, ASI, CoW, Tena]), a domain that is problematic for programmable solutions (i.e., embedded processors) is investigated.

6.1 Benchmarking the RISPP Approach for Different Architectural Parameters

All benchmarks that are presented in this chapter are obtained from the simulation environment that was developed in the scope of the presented work (see Appendix A). It is parameterized using the implementation results from the hardware prototype (e.g., the atom reconfiguration time) and allows to investigate different processor types (ASIPs and different reconfigurable processors) and different architectural parameters [BSH09a]. Table 6.1 summarizes the architectural parameters that were investigated for the RISPP approach.

Figure 6.1 shows the execution time of the H.264 video encoder (processing 20 frames in QCIF resolution, i.e., 176×144 pixels) for different AC quantities. The impact of different frequencies is investigated here, providing the two memory ports with 128 bit each and a 66 MB/s reconfiguration bandwidth to ensure that the frequency results are not affected by a limited memory and reconfiguration bandwidth. It requires 10.6 s [i.e., 1.89 fps (frames per second)] to execute the application on the corresponding GPP (i.e., a Leon2 without hardware accelerators) at 100 MHz and 2.1 s (i.e., 9.52 fps) at 500 MHz. When targeting 30 fps, then the encoding time requires being faster than 0.67 s to fulfill the time constraints. It is noticeable in the figure that there is a significant change in the performance improvement after five ACs are provided to the system. The reason is that for four or less ACs, not all SIs can be implemented with hardware support because insufficient ACs are available. Averaging over all frequencies, shown in Fig. 6.1, there is a 2.00× speedup when providing five ACs instead of four, but only a 1.20× speedup when providing six ACs instead of five. Therefore, the amount of five ACs

Table 6.1 Investigated architectural parameters

Parameter	Symbol	Physical unit
Frequency of the core pipeline	f_{core}	(MHz)
Frequency of the atom infrastructure	f_{atom}	(MHz)
Reconfiguration bandwidth	R	(MB/s)
Number of memory ports	P	N/A
Bit width per memory port	W	(Bits)

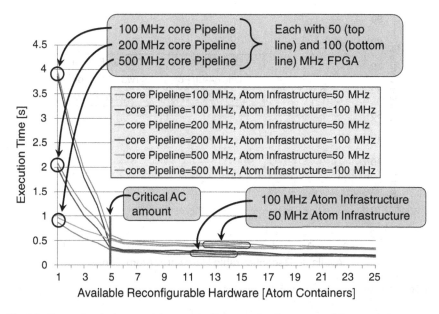

Fig. 6.1 Comparing the impact of the core pipeline operating frequency and the atom infrastructure operating frequency

corresponds to a critical amount. However, providing 15 ACs instead of five still leads to a 1.47× speedup and providing 25 ACs instead of five leads to 1.77×. In comparison to the GPP (averaged over all three core pipeline frequencies), a speedup of 15.91× and 23.56× is noted with 5 and 25 ACs, respectively.

When less ACs than the critical amount are provided, then the core pipeline frequency has a noticeable impact on the overall execution time, and for a sufficient amount of ACs the atom infrastructure has the larger impact. For instance, for f_{atom} = 100 MHZ, changing f_{core} from 100 to 500 MHz results in 3.28× performance improvement when utilizing three ACs but only 1.23× for 20 ACs. Instead, for f_{core} = 500 MHz, changing f_{atom} from 50 to 100 MHz results in 1.09× for 3 ACs, but 1.86× for 20 ACs. As long as not all major SIs are covered by a hardware implementation, some SIs use the cISA execution, i.e., they are executed rather sequential on the core pipeline. Therefore, the frequency of the core pipeline has the larger impact. When more than the critical amount of ACs is available, then the SIs are executed in a parallel manner in the atom infrastructure and therefore the atom frequency becomes more relevant.

For ease of discussion, all following results are presented for 100 MHz core pipeline and 100 MHz for atom infrastructure, i.e., all execution time results are presented in "cycles" instead of "seconds" (only demanded when comparing different frequencies). However, as demonstrated in Fig. 6.1, a faster core pipeline (e.g., 200 or 500 MHz as it can be expected in a nonreconfigurable ASIC implementation of the core pipeline) would not change the results significantly when at least the critical amount of ACs is provided.

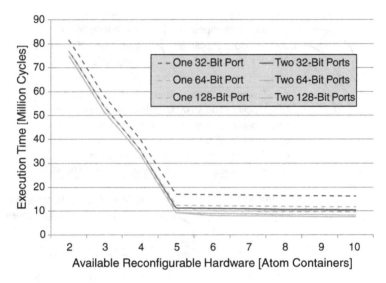

Fig. 6.2 Investigating the effect of different data memory connections

The impact of different data memory bandwidths (that is determined by the amount of independent memory access ports and the bit width per port) is evaluated. In Fig. 6.2, it is noticeable that the settings with two ports (e.g., 2 × 32 bits) always outperform the corresponding single-port settings (e.g., 1 × 64, i.e., providing same total bit width). Improving from 1 × 32 to 2 × 32 results in a 1.68× average speedup, but improving from 2 × 64 to 2 × 128 only achieves 1.15×. Providing two independent ports has the advantage that data streaming can be performed by reading (new input data) and writing (results of previous computations) in parallel. With one port, these operations have to be performed sequentially. It is also noticeable that the number of provided ACs restricts the performance when less than the critical amount of ACs (i.e., five) is available. Even when only one 32-bit port is available, the amount of ACs limits the potential parallelism that can be provided by this amount of input/ output data. After the critical amount of ACs is available, the input data bandwidth limits the potential parallelism.

Finally, the impact of different reconfiguration bandwidths on the overall performance is investigated. Figure 6.3 shows how the execution time decreases for increased bandwidth. For clarity, the readings for less than five ACs (i.e., the critical amount) are omitted, because in this configuration, performance is bound by the number of ACs and the reconfiguration has minimal or no impact. Instead, the importance of the bandwidth when all other bottlenecks (e.g., insufficient input data or insufficient ACs, etc.) are removed (i.e., two 128-bit ports are used here) shall be investigated. As can be seen in the figure, rather slow reconfiguration bandwidth (1–15 MB/s) can be compensated by a higher amount of ACs. As extreme case, if sufficient ACs would be available such that all SIs can be implemented in their largest molecule in parallel, then no further reconfigurations would be demanded

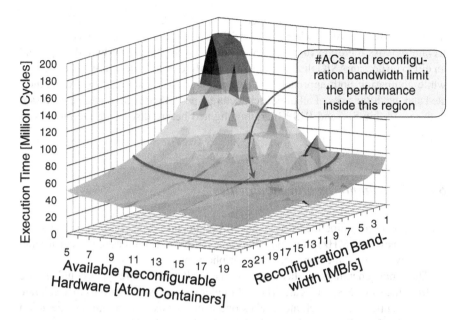

Fig. 6.3 Impact of the reconfiguration bandwidth and the number of atom containers

and thus the reconfiguration bandwidth would not be relevant any more. Vice versa, a low amount of ACs (5–15) can be compensated by a faster reconfiguration bandwidth. As extreme, if the reconfiguration of an atom would only demand one cycle, then an AC could be reconfigured in between two subsequent SI executions and thus the maximum number of ACs required will be the same as required by the largest SI. However, both extremes are unrealistic, but they explain the reason why the number of ACs and the reconfiguration bandwidth may be traded against each other for a given performance constraint. For 25 ACs, increasing the bandwidth from 25 to 66 MB/s only leads to a speedup of 1.06×. However, for ten ACs a speedup of 1.26× can be obtained when increasing to 66 MB/s.

6.2 Comparing Different Architectures

It is not always straightforward to compare different architectures with each other, especially in the case of different concepts and different technologies. This section discusses how the performance comparisons of ASIPs (chooses an implementation alternative during design time) and reconfigurable processors with monolithic SIs (chooses an implementation alternative at compile time) with RISPP (chooses an implementation alternative at run time) are performed, before presenting and analyzing the respective results in Sects. 6.3 and 6.4 in detail. The proposed RISPP architecture is meant to be an ASIC (for the core pipeline, run-time system, and atom infrastructure) with an embedded FPGA (eFPGA) [NvSBN08, SNBN06] for the reconfigurable ACs, whereas an ASIP is completely implemented in ASIC technology. As there are fundamental differences between a hardwired ASIC and an eFPGA (i.e., an ASIC that implements reconfigurable logic), care has to be take while comparing both approaches. The assumptions and similarities are enumerated and the differences of all three architecture types (ASIP, reconfigurable processor with monolithic SIs, and RISPP) are highlighted before discussing the fairness of comparison.

6.2.1 Assumptions and Similarities

1. The same core pipeline is used for all architecture types running at the same frequency (100 MHz) along with the same hardware resources (e.g., register file, read/write ports, memory accesses, periphery, etc.).
2. The same benchmark application is used and all architecture types could choose from the same SIs, molecules (i.e., SI implementations), and atoms to accelerate its execution time. Each atom takes one cycle for execution and the execution time for the different molecules is determined accordingly.
3. RISPP and reconfigurable processors with monolithic SIs additionally use an embedded FPGA that can be reconfigured to contain different atoms or SIs, respectively. For the reconfiguration time of an atom, the implementation results of the FPGA-based prototype (0.70–0.91 ms, see Table 5.7) are considered, and the reconfiguration time of a molecule corresponds to the accumulated reconfiguration times of the demanded atoms.
4. The atoms are considered as basic area unit, i.e., the provided area is represented as the number of atoms/atom containers. This allows a more general comparison that is not implementation specific. For instance, in an early Virtex-II[1]-based prototype, the atom containers had significantly different outlines due to technical constraints[2] of that FPGA family, which lead to different area requirements in comparison to the current Virtex-4 prototype (implementing the same atoms). For the Virtex-5 family, the technical constraints[3] changed again, affecting the area requirements of atoms and ACs again. Therefore, different prototype platforms suggest different area requirement; however, this is technology specific and not concept specific. Instead, using atoms and ACs as elementary area units abstracts from these technology-specific details and allows providing a more general comparison.

---

[1] A Virtex-II 3000 and 6000 were used to develop the first RISPP prototypes; however, this thesis only presents the results for the final prototype that uses a Virtex-4 LX 160.

[2] The smallest run-time reconfigurable element – a frame – spans the full FPGA height for Virtex-II devices.

[3] The LUTs have six inputs instead of four (Virtex-II and Virtex-4) and a frame spans 20 CLBs (16 for Virtex-4).

6.2.2   Dissimilarities

5. ASIPs may achieve a higher clock frequency in comparison to reconfigurable processors and RISPP, because the core pipeline and the atoms are both implemented in ASIC technology. This is also the case for the core pipeline, the run-time system, and the atom infrastructure of RISPP, but the atoms for RISPP and reconfigurable processors are implemented in a potentially slower reconfigurable fabric. However, the major bottleneck for the SI execution is not the frequency, but the limited memory bandwidth (e.g., the SATD SI in Fig. 3.3 requires eight 32-bit inputs and SAD in Table 3.2 demands 64 32-bit inputs). For instance, at 100 MHz frequency, a single-cycle memory access is possible, whereas at higher frequencies the memory access latency (in number of cycles) increases. In addition, the benefit of data caches (that might diminish increased memory access latency) is limited, because the data is typically not needed again. This is due to the streaming nature and the nonlinear access pattern of, e.g., the SAD computation (i.e., a 2D subarray is read and each data word is typically just used once). Therefore, the performance of the SIs (and thus the major computational kernels) does not linearly scale with increase in frequency. Rather, the sequentially executed code that runs on the core pipeline is expected to benefit from an increased frequency, but as the SIs cover all major computational kernels, these parts do not contribute to the overall application execution time significantly.
6. As the atoms for the ASIP are implemented in ASIC technology, they will require less area than those implemented in a reconfigurable fabric. However, the use of embedded FPGAs (e.g., [NvSBN08, SNBN06]), optimized for the proposed ACs, leads to smaller implementations compared with the current prototype [see also assumption (4)], i.e., this effect will diminish. The prototype FPGA is not designed for implementing ACs with their interconnections. Typically, the routing resources of the FPGA become a prototype-specific bottleneck. This is because many of the potentially available routing resources in the area of an AC cannot be used to implement an atom, as they are leaving/entering the AC (instead of staying inside) and may be used by the static design that routes "through" the AC [LBM+06]. For an optimized ASIC for RISPP (as it also has to be created for an ASIP), the ACs would be designed in a more tailor-made way, thus closing the implementation gap to ASIP.
7. As the ASIP atoms are implemented in ASIC technology, they cannot be reused for different operations and thus they typically cannot be used for other than the initially targeted application. Therefore, the ASIP has to offer atoms for all considered applications statically to accelerate all of them. For an increased amount of applications, the size of the ASIP may grow continuously. For instance, the presented H.264 video encoder is just one part of the H.324 video conferencing application (video en/decoder, audio en/decoder, multiplexer and demultiplexer, remote control, and modem interface) that executes besides other applications like encryption or base-band processing. In addition, the highest achievable

frequency of the ASIP may be reduced due to the increased distance between the large amount of atoms and the core pipeline. Instead, RISPP and reconfigurable processors can restrict to a smaller amount of reconfigurable containers (for atoms in case of RISPP and for SIs in case of reconfigurable processors) and then reconfigure them toward the operations that are required at a specific point in time.

8. For benchmarking, the SI implementations (and thus the atoms) are selected in an optimal way (considering the performance) for ASIP and for reconfigurable processor with monolithic SIs, i.e., it was made sure that they get those atoms that lead to the best performance according to their architecture. This does not only comprise exact knowledge of the application, but also exact knowledge of the input video sequence and thus the resulting SI execution frequencies. For (I) HT_4 × 4, MC_Hz_4, IPred_VDC, and IPred_HDC (see Table 3.2), the SI execution frequency depends on the control flow (see the "then" and "else" part in Fig. 3.5). As in real-world scenarios, the motion in the input data cannot be predetermined, ASIP and reconfigurable processor may not have the optimal atoms available for a particular frame (however, the best combination for the overall video sequence is provided), but it will have to encode that frame with a suboptimal set of atoms and thus with suboptimal performance. Instead, RISPP dynamically adapts the SI implementation to cover the current type of motion in the input video sequence (using the online monitoring, forecasting, and molecule selection of its run-time system).

9. The RISPP architecture demands a run-time system (see Chap. 4) to determine the reconfiguration decisions. Even though this achieves dynamic adaptation, it comes at the cost of a static area overhead. Implementing the synchronous part of the run-time system (online monitoring and forecasting, see Sect. 4.3) does not affect the area requirements significantly (see Table 4.1). However, implementing the asynchronous part of it results in additional area requirements. For prototyping, that part is implemented using a dedicated processor (see Sect. 5.5). Note, in a final ASIC with eFPGA implementation, the algorithms of the run-time system should be implemented in an optimized nonreconfigurable design and thus the area requirements will benefit from ASIC technology.

6.2.3 Fairness of Comparison

The fairness of the cross-architecture comparison between ASIPs and reconfigurable processors with monolithic SIs and RISPP is discussed. On one hand, the ASIPs seems to be underestimated, as the atoms could be implemented in better technology [see (5) and (6)] and they do not need a run-time system [see (9)]. On the other hand, RISPP and reconfigurable processors seem to be underestimated as well, as they facilitate a significantly higher flexibility to support different applications. In addition, in comparison to reconfigurable processors with monolithic SIs, RISPP provides further flexibility without demanding the predetermined knowledge of which applications will be executed on which input data pattern [see (7) and (8)].

Depending on the actual requirements and targets objective, some of these points might dominate the others. For instance, if the target application demands few and small SIs, then an ASIP may be privileged. The comparison partner of the ASIP in this case should actually be a dedicated ASIC implementation and not RISPP or a reconfigurable processor. However, consider, for instance, a target system like a mobile device where multiple applications have to be executed (over time and in multitasking) and the owner of the device can download and execute further applications on demand. In these cases, the provided flexibility and dynamic adaptivity of the RISPP architecture will dominate the advantages of a tailor-made ASIP implementation, which can only cover a certain subset of the applications. This is because the ASIP does not scale with an increasing amount of target applications and it cannot address (at design time) unknown applications at all. Such a multitasking scenario is also a critical situation for reconfigurable processors with monolithic SIs. As it is not known during compile time (when these architectures decide about their SI implementation) which tasks will execute together, it is also not known how much reconfigurable fabric is available for a particular SI of an application (as the fabric needs to be shared with other applications). If insufficient fabric is available to reconfigure an SI implementation, then the reconfigurable processor with monolithic SIs has to use the cISA execution for that SI for the entire application execution (until some other tasks release some parts of the reconfigurable fabric), which leads to significant performance degradation.

A processor that is fabricated according to the RISPP architecture does not need to be refabricated when facing different applications, i.e., the RISPP approach is more applicable to different applications and requirement scenarios. When targeting three different applications (e.g., video encoding, encryption, and communication) then three different ASIPs needed to be fabricated, on the contrary, one fabricated RISPP architecture might handle all of them. Therefore, RISPP provides reduced nonrecurring engineering cost in comparison to ASIPs, which might be invested into an improved fabrication technology. Thus, the potential area and frequency advantages of ASIC-implemented atoms for ASIPs [see (5) and (6)] might diminish (depending on the budget and expected selling volume). A reconfigurable processor with monolithic SIs could be retargeted toward all above-mentioned three different applications. However, if they shall execute in a multitasking environment that comes with uncertainty, which applications will execute at which time, and that demands frequent reconfigurations, then the reconfigurable processor cannot retarget the SI implementations without a recompilation/resynthesis to efficiently support changing requirements.

6.2.4 Summary of Comparing Different Architectures

For the core pipeline, RISPP and reconfigurable processors may use the same fabrication technology as the ASIP and the main difference comes in the atoms. It will result in a different atom performance for ASIPs and RISPP, but this highly depends on which ASIP and RISPP technologies are benchmarked and thus it depends on the specific system requirements. Therefore, the aim is to achieve a neutral (and thus general) comparison by considering the cycle-count and the atoms as performance and area measurement unit, respectively. In general, when the particular application or application domain is well known during design time and hardware requirements to accelerate it are moderate, then an ASIP may perform reasonable. If the application domain is not fixed but it is known at compile time which share of the reconfigurable fabric is available for an application (e.g., because only one application shall execute and thus it receives the entire reconfigurable fabric), then a reconfigurable processor with monolithic SIs may be used. However, if a rather large amount of SIs demand frequent reconfigurations or more flexibility is required, e.g., because the computation requirements of the application changes depending on input data or because the reconfigurable fabric is shared among multiple applications, then the superior flexibility of the RISPP approach outperforms state-of-the-art approaches.

6.3 Comparing RISPP with Application-Specific Instruction Set Processors

At first, the performance and execution behavior of an ASIP is analyzed. Figure 6.4 shows the execution time (bars) of the H.264 video encoder for different quantities of deployed atoms. In addition, the efficiency of the atom usages is analyzed, i.e., the speedup per available atom, where the speedup is relative to the execution

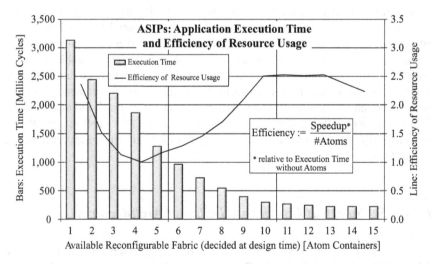

Fig. 6.4 Analyzing the execution time and the resource usage efficiency using different area deployments when processing 140 video frames with an ASIP

Table 6.2 Selected atoms for ASIPs for the computational blocks of H.264 video encoder

Number of used atoms	Selected atoms
1	SADrow
2	SADrow, HT Transform
3	SADrow, HT Transform, Cond
4	SADrow, HT Transform, Repack, QSub
5	SADrow, HT Transform, Repack, QSub, SAV
6	SADrow, HT Transform, Repack, QSub, SAV, DCT Transform
7	SADrow, HT Transform, Repack, QSub, SAV, DCT Transform, Cond

time of the core pipeline without any hardware accelerators (requiring 7.4 billion cycles to encode 140 video frames with an H.264 encoder), see Eq. 6.1.

$$\text{Efficiency} = \frac{\text{Speedup*}}{\#\text{Atoms}}.$$
*Relative to the execution time without atoms. (6.1)

Table 6.2 shows the selected atoms for the first seven readings in Fig. 6.4. Note that some of the selected atoms are reused to implement different SIs, e.g., HT Transform is used by SATD and (I)HT_4 × 4. One interesting situation can be seen when moving from three to four atoms in Table 6.2. While in each other increment step, the previously selected atoms are extended by an additional atom, in this step the previously selected Cond atom is discarded and two new atoms are selected instead. This is because Repack and QSub are required together to achieve a notice-able performance improvement of SATD, whereas LF_BS4 can be accelerated even if Cond is the only available atom.

An SI can be implemented with a subset of its required atoms, e.g., the SATD SI can be accelerated (with a smaller speedup) even if only the HT Transform atom is available (the remaining computation is performed without hardware accelerators). Therefore, the speedup for the first added atom is already 2.4×, which results in a good efficiency (see the efficiency line in Fig. 6.4). However, to achieve a better performance, more atoms have to be added. This leads to a significant efficiency decrease, e.g., doubling the number of atoms does not double the speedup. This is because only few SIs benefit from the small amount of atoms. To obtain a good compromise between execution time and efficiency in Fig. 6.4, at least ten atoms have to be used. After adding 13 atoms, the efficiency is again decreasing, as the speedup is limited by the sequential part of the application. The result of this analysis is that up to a certain quantity of atoms, the resource utilization of an ASIP is ineffi-cient, thus limiting the potential speedup. A rather large amount of atoms need to be added to match the required performance. This is a significant problem considering large applications like H.324 video conferencing (the H.264 video encoder is one component of H.324) or even multiple applications in a multitasking environment.

To understand the underlying reason of this inefficient resource utilization, the SI execution pattern and the corresponding atom usages is examined. Figure 6.5 illustrates the problem by showing the detailed atom utilization for timeframes of

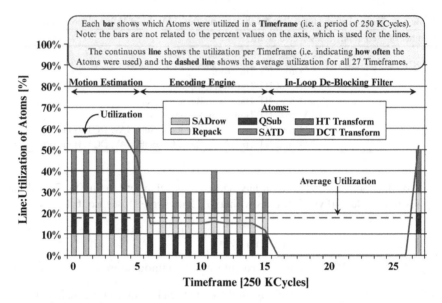

Fig. 6.5 Detailed ASIP utilization variations for six available atoms

250 K cycles (*X*-axis) for six available atoms (corresponding to the atoms shown in Table 6.2). The atom utilization is defined as shown in Eq. 6.2. This definition is based on the observation that in the best case, each SI execution in a timeframe could make use of all available atoms, which then corresponds to 100% utilization according to the definition. The bars in Fig. 6.5 show which atoms were actually used per timeframe (for clarity it is not shown how often they were used). The drawn-through line corresponds to the atom utilization and the dashed line shows the average atom utilization for the whole execution. The maximum number of available atoms is only used in timeframe 5. In this timeframe, the processing flow changes from motion estimation (using five atoms) to encoding engine (using three atoms), i.e., not all six atoms are used for the same computational block. It can be seen that the Repack, QSub, and HT Transform atoms are used for both computational blocks, thereby increasing the average utilization and thus the overall efficiency. Except these three, all other atoms are dedicated to a specific computational block and therefore they are not utilized efficiently. In timeframes 16–26, (execution of in-loop de-blocking filter) not even one of the available atoms can be used, which results in a disadvantageous average utilization of 17.7%. This is because no atom was selected for the loop filter SI when at most six atoms may be available. Instead, to achieve the best overall performance, all six available atoms were given to the motion estimation and encoding engine.

$$\text{Atom utilization} = \frac{\#\,\text{Executed SIs} \times \#\,\text{Available atoms}}{\#\,\text{Actually used atoms}}. \qquad (6.2)$$

This analysis shows that the efficiency of ASIPs is rather moderate when only a few atoms are provided. However, to reach a good operation point with a high efficiency, many atoms would need to be added. The small speedup when only a few accelerating atoms are provided is due to an inefficient utilization of the available hardware resources. If the efficiency of the hardware usage can be improved, then a good performance can be achieved with fewer atoms. Thus, area would be saved (and costs, static power consumption, etc.) or the area could be used for other components, e.g., caches. This problem is tackled by the RISPP approach that uses the available hardware for atoms in a time-multiplexed way to implement modular SIs. When the processing of one computational block is completed, the atoms are reallocated to the SIs of the subsequent computational block.

Figure 6.6 depicts the comparison of an ASIP and RISPP for different amounts of ACs (reconfigurable for RISPP and nonreconfigurable for ASIP), respectively. This analysis shows the execution time (bars: left Y-axis) and efficiency (lines: right Y-axis) of the benchmark application for ASIP and RISPP. RISPP is up to 25.7 (average 17.6) times faster than the GPP (using no hardware accelerators) and up to 3.1 (average 1.4) times faster than the ASIP (all executing at 100 MHz as discussed in Sect. 6.2). The execution time of the application using a GPP without hardware accelerators is 7.4 billion cycles. It can be noticed from the figure that for five ACs, RISPP has the maximum efficiency of 3.6 (this is 3.0 times better than that of ASIP using the same amount of ACs) as it can already execute all SIs in hardware (due to the time-multiplexed hardware usage, i.e., reconfiguration). The execution time of the application in that case is 416.67 million cycles, i.e., 17.8 times better performance than the GPP.

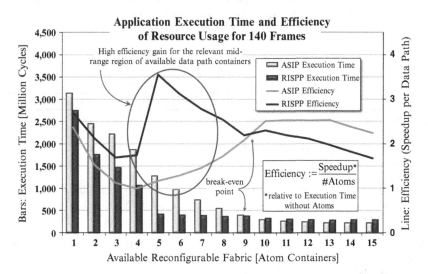

Fig. 6.6 Application execution time and efficiency of resource usage for encoding 140 video frames on ASIP and RISPP

When using up to nine ACs, the performance of RISPP is better than that of the ASIP. This point is called "break-even point," as both architectures have nearly the same efficiency and execution time because all SIs are executed completely in hardware. Beyond the break-even point (i.e., when providing more ACs), the reconfiguration delay dominates the performance gain of faster hardware implementations of SIs. ASIP starts winning beyond the break-even point, but after some further addition, performance saturation is reached (due to the available parallelism in the application, i.e., Amdahl's law). RISPP instead achieves already a good performance with a high efficiency for a rather small number of ACs. Table 6.3 summarizes the important attributes of the comparison of RISPP and ASIP. In comparison with a GPP, the ASIP achieves the higher maximum speedup (33.6× in comparison to 25.7× for RISPP, both using 15 ACs). The reason is that – given a rather large amount of atoms to the ASIP – all SIs can be accelerated in hardware even without reconfigurations. In this scenario, the ASIP achieves its highest speedup, however, at significant hardware cost. The comparison of the "efficiency" (i.e., speedup per atom) in Table 6.3 reflects these hardware costs. The minimum, average, and maximum efficiency of RISPP is better compared with the ASIP. In addition, the minimum and average speedup of RISPP in comparison to GPP is better than comparing ASIP with GPP. The ASIP only achieves the better maximal speedup, if a large amount of atoms is available. When directly comparing RISPP with the ASIP, then the ASIP peak performance corresponds to a performance loss of 0.75 for RISPP. However, as RISPP provides the better performance for the majority of benchmark situations (i.e., availability of ACs), RISPP's average speedup in comparison to the ASIP is 1.38 and even 3.06× speedup is reached for five ACs. When focusing on moderate area extension by providing one up to nine ACs (the break-even point), then the minimum and average speedup of RISPP in comparison to the ASIP improve to 1.05× and 1.75×, respectively (see Table 6.3).

Table 6.3 Summary of comparison of RISPP and ASIP

	ASIP			RISPP		
	Minimum	Average	Maximum	Minimum	Average	Maximum
Execution time (MCycles)	220.6	999.6	3126	288.3	715.1	2,734
Speedup in comparison to GPP	2.4	16.8	33.6	2.7	17.6	25.7
Efficiency	1.0	1.9	2.5	1.7	2.3	3.6
Speedup in comparison to ASIP	–	–	–	0.75	1.38	3.06
Speedup in comparison to ASIP when focusing on 1–9 ACs	–	–	–	1.05	1.75	3.06

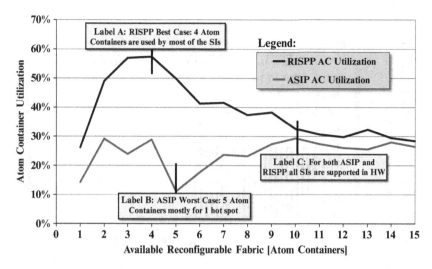

Fig. 6.7 Atom utilization for ASIP and RISPP

Figure 6.7 demonstrates the resource utilization (see Eq. 6.2) of RISPP and ASIP. The figure shows that the resource utilization of RISPP is much better than that of an ASIP when few ACs are available due to the time-multiplexed utilization of the hardware resources (i.e., run-time reconfiguration). Three significant points (see the labels in Fig. 6.7) are analyzed with the help of a corresponding run-time analysis (shown in Figs. 6.8–6.10) in detail. Label A in Fig. 6.7 represents the best case of resource utilization (57.4%) corresponding to four ACs for RISPP. Figure 6.8 shows the detailed atom utilization for this case. In the first computational block (i.e., motion estimation (ME); see Fig. 3.5), all four ACs are used. After ME completed execution, RISPP reconfigures the ACs for the next computational block (i.e., encoding engine), as it can be seen at the temporary drop of the resource utilization (an AC cannot be used during its reconfiguration). In timeframe 23, RISPP uses more than four atoms, as in the beginning of this timeframe the atoms for the ME are still available and used, where at the end of this timeframe the atoms for the encoding engine are already loaded and used. However, at no time more than four atoms are used together as only four ACs are available. In timeframes 29 and 30, the ACs are reconfigured toward the third computational block (i.e., in-loop de-blocking filter), before the execution of the three computational blocks restarts for the next input frame (not shown for clarity).

All figures show the encoding of one video frame. The X-axis shows the timeframes of 250 K cycles, the left Y-axis denotes the overall and average utilization (drawn as lines), whereas the bars indicate which atoms were used in the current timeframe. The length of the X-axis corresponds to the maximum encoding time (for one frame) from ASIP or RISPP (partially showing the beginning of the second frame).

It is noticeable in Fig. 6.8 that the ASIP spends all four atoms for the ME, as this enables the best performance for the overall encoding. In timeframe 34 (within the

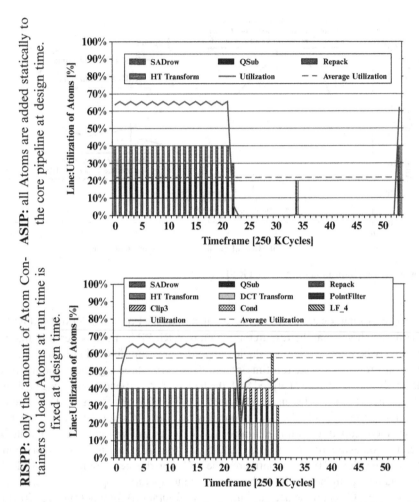

Fig. 6.8 Detailed atom utilization variations for four available atom containers (ACs), used by ASIP and RISPP, respectively

encoding engine), the ASIP reuses the HT Transform and Repack atoms to encode some MacroBlocks using intraframe prediction (I-MBs). RISPP instead reconfigures the ACs to support the more frequent P-MBs and accepts using the cISA execution to encode the I-MBs. By doing so, RISPP achieves an encoding time that is significantly faster than that of the ASIP. RISPP executes ME slightly slower than the ASIP due to the initial reconfiguration overhead, but the encoding engine and loop filter overcome this initial shortfall.

Label B in Fig. 6.7 portrays the worst-case resource utilization (11%) for the ASIP, when it uses five atoms. This is because – in addition to the atoms that were also used when at most four atoms were available – the ASIP additionally offers the

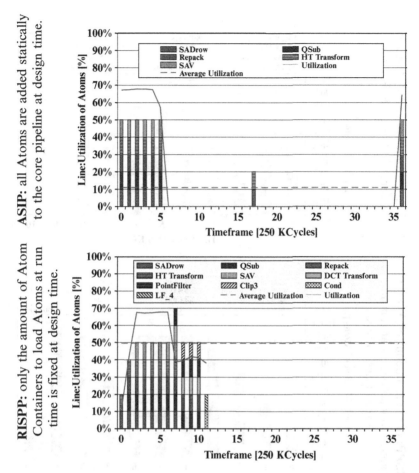

Fig. 6.9 Detailed atom utilization variations for five available atom containers (ACs), used by ASIP and RISPP, respectively

SAV atom (see Fig. 6.9) that is only beneficial for the SATD SI. This leads to the best performance (6 instead of 23 timeframes for ME), but the encoding engine and loop filter do not benefit at all and now dominate the execution time. At Label C in Fig. 6.7, the atom utilization between RISPP and ASIP is similar. Now also for the ASIP, all three computational blocks are covered with hardware-accelerated SIs (although not with the fastest implementations).

Figure 6.10 shows that the RISPP architecture is already affected by the reconfiguration time. The first four timeframes are spent in reconfiguring all selected atoms for the ME computational block, which does not only have a negative impact to the utilization, but also to the performance. The same is true for the encoding engine and partially for the in-loop de-blocking filter. Therefore, only in timeframe 10, all provided ACs are actually used. Finally, the better SI implementations that

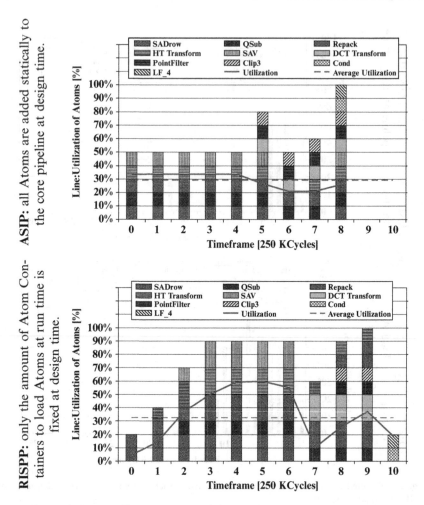

Fig. 6.10 Detailed atom utilization variations for ten available atom containers (ACs), used by ASIP and RISPP, respectively

RISPP selected no longer overcome the reconfiguration overhead. An improved reconfiguration time would attenuate this effect and improve the performance and atom utilization of the RISPP architecture even further.

6.4 Comparing RISPP with Reconfigurable Processors

In addition to the comparison with GPPs and ASIPs, this monograph also presents a comparison of RISPP with state-of-the-art reconfigurable processors with monolithic SIs. In Sect. 2.2.4, e.g., Molen [VWG+04] and OneChip [WC96] were identified as state-of-the-art approaches and described accordingly. Even though there are differences in the architecture details between RISPP and them, they represent the class of processors with reconfigurable monolithic SI. For instance, Molen does not call the SIs directly, but data transfer is managed by so-called exchange registers. In addition, Molen performs static prefetching to reconfigure the SIs, i.e., at compile time it is predetermined which implementations shall be loaded. To simulate static prefetching with the simulation environment (see Appendix A), the fine-tuning operations for forecast values are turned off by setting α to zero (see Sect. 4.3.1). This assures that always the same reconfiguration decisions are derived. In addition, only one implementation is provided per SI, which represents the fact that Molen statically provides one reconfigurable accelerators at compile time (when they are synthesized). To assure a fair comparison, the same SIs, molecules, and atoms are provided to Molen (see discussion in Sect. 6.2). For each size of the reconfigurable fabric (i.e., number of available ACs), those molecules were selected that together lead to the best performance according to the exact SI execution frequencies (i.e., the profiling results for the video sequence were used to determine the implementation, although the actual video sequence is typically not known when compiling the application). Therefore, this comparison corresponds to a conservative comparison, as it uses information that is typically not available to optimize statically for the targeted video sequence.

As Molen provides exactly one dedicated implementation per SI, the simulation assured that the atoms that are used for a particular SI may not be used to implement any other SI, i.e., atom sharing – one of the novel concepts of the RISPP architecture – is not available for Molen. However, when a new SI shall be loaded for Molen, it is reconfigured at the basis of atoms, i.e., the area unit that is used for comparison. If none of the demanded atoms are available, then the reconfiguration time corresponds to the reconfiguration time of a monolithic SI. However, it may happen that some of the demanded atoms are available before the reconfigurations start, because only one atom is replaced, if a new one is loaded. In case of monolithic SIs, either the entire SI implementation is replaced or none of it, whereas in the simulation only the fraction of the SI implementation (i.e., the atom) that is actually overwritten is no longer available. All other atoms stay available and do not need to be reconfigured for the next time this SI is demanded. Molen actually does not provide this performance-increasing feature and thus, providing it to Molen, leads to a conservative comparison again. In addition, the simulation allows that the implementations of different SIs differ in their size, i.e., an SI may be implemented with rather few atoms, whereas another SI is implemented with rather many atoms. However, for monolithic SIs – as used by Molen – the reconfigurable fabric is typically partitioned into identical-sized regions into which SI implementations can be loaded. This is done to allow any SI implementation to be loaded into any of these slots (same reasons, why ACs have all the same size and shape). Eventually, the simulation assures that the Molen simulation does not used forecast fine-tuning, adaptive molecule selection, SI upgrading, and atom sharing, which are dedicated features of the novel RISPP approach that is presented in this monograph.

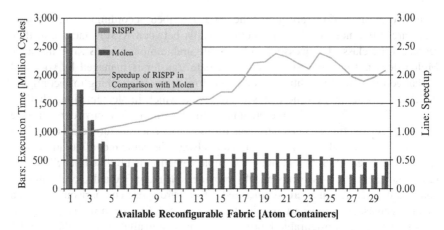

Fig. 6.11 Comparison of RISPP and Molen, showing the execution time of the h.264 video encoder benchmark (encoding 140 frames in CIF resolution, i.e., 352 × 288) and the speedup

Figure 6.11 shows the comparison of RISPP and Molen for different sizes of the reconfigurable fabric. The bars compare the corresponding application execution times, and the line shows the resulting speedup of RISPP in comparison to Molen. RISPP achieves up to 2.38× (for 24 ACs) speedup. In average, RISPP achieves 1.69× speedup compared with Molen, and it is noteworthy that it never performed slower than Molen (see Table 6.4. This shows that the RISPP concept is superior to state-of-the-art reconfigurable processors where special instructions (SIs) are determined at compile time and are fixed at run time, whereas RISPP can gradually upgrade depending on the state of the system. It is noticeable, that Molen performs reasonably well, when rather few ACs are available. A small reconfigurable fabric leads to correspondingly small SI implementations (i.e., consisting of few atoms) and thus the reconfiguration time is not a major issue here. However, when a larger reconfigurable fabric is available, then the reconfiguration overhead affects the performance significantly (as also discussed when comparing RISPP with ASIPs in Sect. 6.3). RISPP's concept of modular SIs diminishes this affect, as it becomes visible in Fig. 6.11. It is interesting to see that for nine ACs, the performance of Molen is viewable worse than for eight ACs (due to reconfiguration overhead).[4] A reconfigurable fabric with nine ACs corresponds to the break-even point where the ASIP afterwards (i.e., for ten ACs) started providing a faster performance in comparison to RISPP (see Fig. 6.6). The same effect (reconfiguration overhead)

---

[4]Please note that the SI implementations for Molen were selected optimally, considering how the reconfigurable fabric should be shared among the SIs (that are demanded in the computational blocks of the benchmark application) and considering the actual SI execution frequencies, but not considering "when" and "how often" Molen might reconfigure the SIs. When provided with nine ACs, a Molen compiler may decide to use only eight of them; however, still Molen's performance does not improve beyond six ACs, whereas RISPP achieves noticeably better performance when equipped with more ACs, providing up to 1.94× speedup in comparison to Molen's fastest execution time.

Table 6.4 Speedup compared with Molen, a state-of-the-art reconfigurable processor with monolithic SIs

	Minimum	Average	Maximum
Speedup of RISPP in comparison to Molen	1.00	1.69	2.38
Speedup of RISPP in comparison to Molen when focusing on 10–30 ACs	1.31	1.94	2.38

affects RISPP and Molen, but RISPP handles the situation much better (leading to a better performance) due to its novel concept of modular SIs. Therefore, in Table 6.4, also the speedup of RISPP for a relatively large reconfigurable fabric of 10–30 ACs is presented (i.e., after the break-even point). After 22 ACs, the performance of Molen recovers primarily because in the Molen simulation, the atoms that are not explicitly replaced but they remain loaded in the reconfigurable fabric, even though the monolithic SIs is replaced. As discussed beforehand, Molen actually does not provide this feature, but to use the ACs as area unit that feature is provided to Molen as well to maintain a conservative comparison.

Up to now, the benchmarks are performed for situations where it is clear at compile time which amount of reconfigurable fabric (i.e., number of ACs) is available to execute the SIs that the application demands, and the SI implementations are optimally selected according to this information by considering the SI execution frequencies. However, the size of the reconfigurable fabric may be neither known (at compile time) nor constant (during run time). For instance, in case of a multitasking system where multiple tasks may demand a share of the reconfigurable fabric, the amount of ACs that may be assigned to a particular application may change (e.g., when new applications are started or the priorities of the applications change). Proteus [Dal03] is a state-of-the-art reconfigurable processor that uses monolithic SIs and investigated scenarios with explicit support for operating systems. Therefore, RISPP is compared with Proteus, when facing the effects of run-time changing availability of reconfigurable fabric.

Proteus uses multiple reconfigurable functional units (RFUs) and may reconfigure them to contain entire SI implementations. As up to four atoms are demanded to implement one SI in the smallest hardware implementation (SATD, see Fig. 3.3), one RFU is represented as four ACs for simulations. Figure 6.12 provides an example that shows the conceptual problem when statically deciding about the SI implementations. Figure 6.12a shows the Proteus, using monolithic SIs and providing three RFUs. Please note that only those RFUs are shown that are assigned to the application, potentially further RFUs might be available in the system. In the example, three SIs from the encoding engine computational block of an H.264 video encoder are loaded into the RFUs. If the operating system (OS) reassigns a fraction of the reconfigurable fabric that was previously assigned to the video encoder, then the smallest unit that can be reassigned is a RFU. Therefore, only two SIs of the encoding engine are accelerated by hardware, which results in a significant performance degradation. Here, it would be better to repartition the reconfigurable

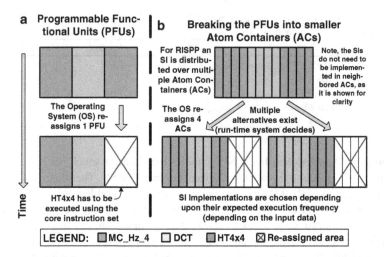

Fig. 6.12 Problem and possible solution when addressing reconfigurable hardware as monolithic RFUs

fabric that is assigned to the application to provide three smaller RFUs. Then, all three SIs could still be accelerated by hardware (though exploiting less parallelism compared with the SI implementations for the larger RFUs) which removes the major performance bottleneck, i.e., executing HT4 × 4 using the cISA as shown in Fig. 6.12a. The RISPP approach shows multiple advantages here due to the concept of modular SIs and ACs. At first, the smaller ACs provide the opportunity for the OS to distribute the available reconfigurable fabric at a finer granularity. Furthermore, even if the same amount of reconfigurable fabric is reassigned (corresponding to four ACs as shown in Fig. 6.12b) still all SIs can be implemented using hardware accelerators, i.e., atoms. In addition, multiple alternatives exist how the ACs can be used to implement the SIs, whereas it is also possible that some atoms that are loaded into the ACs are shared by different SI implementations (not shown in the figure).

In Fig. 6.13, the performance of RISPP and Proteus is compared. It is worth to note that the performance improvement (when providing more reconfigurable fabric) for Proteus is discontinuous. The reason is that the system is optimized to RFUs that correspond to an area equivalent of four ACs (as four atoms are demanded to implement SATD as discussed beforehand). This decision was done at compile time without knowing how many RFUs would be available at run time (due to multitasking). For the comparison with Molen, the SI implementations were optimized specifically for the amount of ACs that were available at run time (as Molen does not target multitasking systems). For Proteus, the RFU size is decided once and then has to be used for all different scenarios, i.e., number of available RFUs. If bigger RFUs had been used, then the above discussed effect of "discontinuous performance changes" would increase, i.e., reassigning the area equivalent of one AC to another task might lead to even larger performance changes. If smaller

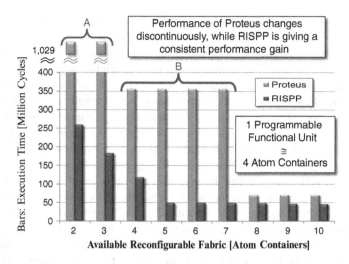

Fig. 6.13 Comparing the RISPP approach with Proteus, executing an H.264 video encoder and showing the impact of RFUs vs. atom containers

RFUs had been used then not all important SIs could be offered in hardware at all (independent of how much reconfigurable fabric is available). In contrast to this, the RISPP approach uses the hardware resources efficiently and gives a consistent performance improvement for each additional AC. Labels A and B in Fig. 6.13 highlight two prominent groups as they correspond to zero and one RFU, respectively. Altogether, RISPP is up to 7.19× (on average 3.63×) faster than Proteus. In group B, RISPP is on average 6.14× faster. After group B, the difference is rather less, as both RISPP and Proteus are now offering the most frequent SIs in hardware. However, consider that the available reconfigurable fabric has to be shared between multiple tasks, and the figure only shows which share of the reconfigurable fabric may be used for this particular application.

6.5 Summary of Benchmarks and Comparisons

This chapter benchmarked the novel RISPP approach that is presented in this monograph, using different scenarios.[5] The impact of different frequencies was investigated for the core pipeline and the atom infrastructure (containing the reconfigurable fabric), depending on the amount of available atom containers (ACs). In addition, the RISPP approach was investigated for different memory bandwidths (number of ports and bit width per port) and reconfiguration bandwidth.

---

[5] Please note that specific benchmark results for the individual components of the RISPP run-time system and the RISPP architecture prototype are presented in Chap. 4 and Sect. 5.5, respectively.

The results show that the parameters interact with each other, and the impact of setting a parameter to a certain value also depends on the settings of the other parameters. For instance, the frequency of the core pipeline is very important if a relatively small amount of reconfigurable fabric is available, whereas it has a small impact when many ACs are available. This is because if sufficient number of ACs are available, then large fraction of computation is performed using them. In such a case, the impact of the data memory bandwidth is important to exploit the parallelism, whereas its importance is limited if rather few ACs are available that could exploit the potential parallelism of a large data memory bandwidth. Similarly, the importance of the reconfiguration bandwidth depends on the amount of available ACs. If relatively few of them are available, then they need to be reconfigured more frequently and thus the reconfiguration bandwidth determines when they can be used (i.e., when they finish the reconfiguration). However, if rather many ACs are available, then some of them may retain their atoms, i.e., they are not reconfigured even though the computational block finished execution. Therefore, these atoms will already be available next time when the computational block is executed and less additional reconfigurations are demanded. Thus, the impact of the reconfiguration bandwidth is not that important as for the case where frequent reconfigurations occur.

In addition to these benchmarks, this monograph also compares the proposed approach with state-of-the-art nonreconfigurable ASIPs and with state-of-the-art reconfigurable processors. When comparing with ASIPs, it becomes apparent that they provide relatively bad efficiency when rather few atoms[6] are available. This is because the atoms are dedicated to a specific functionality and have to address all requirements (i.e., SIs) of the application. As they cannot be reconfigured when the execution moves from one computational block to another, some of them are not used for a significant amount of time (e.g., the execution time of a computational block). Therefore, RISPP achieves an up to 3.06 better performance (1.75× in average) than that of an ASIP that is statically optimized for the specific application and input data. When rather many atoms are available, then an ASPI may also perform better than RISPP (up to 1.33×) due to RISPP's reconfiguration overhead (even though the RISPP approach diminishes this effect). However, RISPP provides significantly increased adaptivity in comparison to an ASIP, i.e., it is not optimized for any particular application. Therefore, in scenarios where it is not known beforehand which applications will execute (e.g., because the user can download and start further applications), the ASIP might perform as bad as a GPP, whereas RISPP can reconfigure toward the demanded application dynamically. In comparison to a GPP, RISPP provides up to 25.7× faster application execution time.

When comparing with state-of-the-art reconfigurable processors, the performance was analyzed for situations where the availability of reconfigurable fabric (i.e., the number of ACs) is known at compile time (as it is typical in single-tasking scenarios) and where it is not known (as it is typical in multitasking scenarios).

---

[6] Actually, their nonreconfigurable counterparts used by the ASIPs.

For the known number of ACs, comparison with Molen [VWG+04] is provided, using statically optimized SI implementations for Molen, targeting the particular scenario. Still, RISPP is able to perform up to 2.38× faster than Molen (in average 1.69× and never slower than Molen) which shows the superiority of the RISPP approach and its modular SIs. Especially when rather many ACs are available, RISPP performs better due to its diminished reconfiguration overhead and adaptivity. For run-time changing availability of reconfigurable fabric, the extended adaptivity of RISPP leads to even larger performance improvements. In comparison to Proteus [Dal03], which explicitly targets multitasking scenarios, RISPP achieves up to 7.19× better performance (in average 3.63× and never slower than Proteus), because RISPP can dynamically select the SI implementations depending on the availability of ACs, whereas Proteus has to use a compile-time determined decision independent of the amount of ACs that are available during run time.

Altogether, these benchmarks and comparisons demonstrate RISPP's superiority in scenarios where rather many SIs demand acceleration, the SI execution frequency depends on input data, or the availability of ACs is not fixed, i.e., in scenarios where a high adaptivity is beneficial. The novel concept of modular SIs and the algorithms of the run-time system that dynamically select SI implementations and determine reconfigurations improve the provided adaptivity in comparison to ASIPs and state-of-the-art reconfigurable processors, and additionally diminish the reconfiguration overhead.

Chapter 7
Conclusion and Outlook

7.1 Summary

This monograph presents a novel approach to increase the adaptivity and efficiency of embedded processors, i.e., it presents RISPP, the rotating instruction set processing platform. RISPP combines the aspects of application-specific instruction set processors (ASIPs) and reconfigurable processors by providing special instructions (SIs) that may be reconfigured during run time, i.e., a run-time adaptive instruction set. The contribution of RISPP is based on the novel concept of modular SIs. Instead of providing an SI implementation as a monolithic block, modular SIs provide multiple different implementations of an SI. An SI is partitioned into multiple elementary data paths, the so-called atoms. These atoms are the basic unit that may be reconfigured to the reconfigurable fabric. The different implementations of an SI vary in their amount of provided atoms and thus correspond to a certain "performance per area" trade-off. In addition, atoms are not necessarily dedicated to a particular SI but their functionality might be used by different SIs and their implementations, i.e., an atom can be shared.

In addition to the novel concept of modular SIs, this monograph presents a novel run-time system that exploits the potential adaptivity and performance benefits. Based on online monitoring and a lightweight prediction scheme, the run-time system determines which implementation of an SI shall be used to accelerate a computational block of an application. This selection provides adaptivity for changing application control flow and changing multitasking scenarios. If the application control-flow changes during run time (e.g., because it depends on input data), then the execution frequency of some SIs might change. In such a situation, the run-time system can provide more atoms to accelerate the corresponding SIs, i.e., it adapts to the compile-time unpredictable situation. In multitasking scenarios, the available reconfigurable fabric needs to be shared among all tasks that shall be accelerated. Depending on the number of tasks and their priorities or deadlines, the amount of available reconfigurable fabric that is dedicated to a particular task is not predetermined. Again, the run-time system allows adapting to these changing multitasking scenarios by selecting correspondingly smaller or larger implementations for the demanded SIs.

L. Bauer and J. Henkel, *Run-time Adaptation for Reconfigurable Embedded Processors*,   191
DOI 10.1007/978-1-4419-7412-9_7, © Springer Science+Business Media, LLC 2011

The novel run-time system also determines the atom loading sequence, as at the most one atom can be loaded at a time. Reconfiguring another atom may allow that – together with the already available atoms – a faster implementation of a particular SI becomes available, i.e., the SI is upgraded. To exploit the performance of modular SIs, these SI upgrades need to be considered carefully, which is achieved by the highest efficiency first scheduler of the presented run-time system. In addition, whenever an atom is reconfigured, another atom might need to be replaced, which is performed by the novel minimum degradation replacement policy, presented in this monograph. The novel feature of stepwise upgrading the implementation of an SI diminishes the conceptual reconfiguration overhead problem of state-of-the-art reconfigurable processors that provide monolithic SI implementations. The reconfiguration time of these monolithic SIs becomes longer if the SI implementation exploits more parallelism (because a larger implementation leads to more configuration data that needs to be transferred). Therefore, the reconfiguration overhead limits the amount of parallelism that can be exploited by monolithic SIs. In extreme case, the reconfiguration could demand more time than the execution of the kernel without hardware acceleration. However, the concept of modular SIs as presented in this monograph allows exploiting the maximally available parallelism, because the SI implementations can be upgraded until the most parallel versions are available. In between, "smaller" implementations (i.e., demanding less atoms and exploiting less parallelism) are available to accelerate the application. To implement modular SIs in practice, this monograph presents a novel computation and communication infrastructure that supports SI upgrading. It is coupled to the pipeline of the core processor like a functional unit and has access to the data memory.

All proposed algorithms for the novel run-time system are described on a formal basis and evaluated for different parameter settings or algorithmic versions. The entire RISPP approach, including the run-time system and the atom infrastructure is implemented on an FPGA-based prototype to allow partial run-time reconfiguration. A challenging H.264 video encoder application is used to benchmark RISPP and evaluate the performance impact of different parameters, i.e., amount of available reconfigurable fabric, frequency of the core pipeline, frequency of the atom infrastructure, available data memory bandwidth, and available reconfiguration bandwidth. In addition, this monograph provides a comparison of RISPP with a general-purpose processor (GPP), a state-of-the-art ASIP, and state-of-the-art reconfigurable processors (Molen [VWG+04] and Proteus [Dal03]). In comparison to a GPP, RISPP provides up to 25.7× faster application execution time, running at the same frequency. RISPP achieves an up to 3.06× better performance (1.75× in average) than that of an ASIP that is statically optimized for the specific application and input data. It is noticeable that RISPP achieves this performance improvement for a rather small reconfigurable fabric as its concept of dynamically reconfiguring modular SIs uses the provided hardware more efficient. When a rather large amount of hardware is available, then the ASIP may potentially implement all SIs in their most parallel version. In this case, the ASIP is 1.33× faster than RISPP, however, at the cost of an explicit specialization for a particular application and a large area footprint. Instead, RISPP may reconfigure the atoms to support different applications as well, i.e., it is not optimized for any particular application, but it is flexible. Therefore, in scenarios where it is not known beforehand which applications will execute (e.g., because the user can download and start further applications), the ASIP might perform as bad as a GPP, whereas RISPP can reconfigure toward the demanded application dynamically.

Comparing with Molen, RISPP achieves a performance improvement up to 2.38× (in average 1.69× and never slower than Molen), which is due to the concept of modular SIs that (a) diminishes the reconfiguration overhead when exploiting SI implementations with a high degree of parallelism and (b) provide adaptivity when facing input-data-dependent SI execution frequencies. These significant improvements are obtained even though Molen was statically optimized for the particular input data that was used for benchmarking (which is not possible in real-world scenarios), whereas RISPP was adapting to the SI execution frequencies dynamically over time. When facing multitasking scenarios with changing availability of the reconfigurable fabric (because it is shared among multiple tasks), then the extended adaptivity of RISPP improves the performance further. In comparison to Proteus [Dal03], which explicitly targets multitasking scenarios, RISPP achieves up to 7.19× better performance (in average 3.63× and never slower than Proteus), because RISPP can dynamically select the SI implementations depending on the availability of ACs, whereas Proteus has to use a compile-time determined decision independent of the amount of ACs that are available during run time.

Altogether, these benchmarks and comparisons demonstrate RISPP's superiority in scenarios where rather many SIs demand acceleration, the SI execution frequency depends on input data, or the availability of ACs is not fixed, i.e., in scenarios where a high adaptivity is beneficial. The novel concept of modular SIs and the algorithms of the run-time system that dynamically select SI implementations and determine reconfigurations improve the provided adaptivity in comparison to ASIPs and state-of-the-art reconfigurable processors, and additionally diminish the reconfiguration overhead.

7.2 Future Work

The conceptual benefits, the various benchmark results, and the comparison with state-of-the-art embedded processors provided in this monograph demonstrate that adaptivity makes it possible to improve the performance and efficiency of today's embedded processors and thus embedded systems. These promising results enable further research efforts that may focus on compile-time automation, broadened run-time adaptivity, and improved design efficiency:

Compile-time automation. In the scope of the work presented in this monograph, the special instructions (SIs), molecules, and atoms were manually created, as stated in Sect. 3.3 and presented in Shafique et al. [SBH09a]. As the design methodology and tool flow for the application designer and the instruction-set architecture are related to ASIP design, the tools and algorithms of that research domain may be adapted to create SIs for RISPP automatically. However, SIs for state-of-the-art ASIPs and reconfigurable processors are monolithic, i.e., they are not partitioned into elementary data paths, i.e., atoms. Therefore, variations of graph partitioning algorithms may be investigated to transform monolithic SI graphs into modular SIs and to determine which properties modular SIs demand. This can be used to modify the "pruning" step in state-of-the-art automatic SI detection (see, for instance, [SRRJ04, VBI07]). To exploit the feature to share atoms between different SIs, techniques like data-path merging [BKS04] may be adapted to identify reusable atoms. The research project KAHRISMA [ITI, KBS+10] started, which – to some degree – builds upon the results of the work that was presented in this monograph. As a part of that project, the automatic detection and partitioning of modular SIs will be investigated further.

Broadened run-time adaptivity. Even though a large adaptivity is already exploited by the approach presented in this monograph, further potential is available when considering energy efficiency. Up to now, RISPP aims to transfer the available conceptual advantages into pure performance. However, given certain deadlines, i.e., performance constraints, the question arises whether the conceptual advantages can be used to fulfill the performance constraints while optimizing for low-energy consumption. For instance, upgrading SIs provides improved performance but costs reconfiguration energy. Recent work investigated this matter [SBH09b], using the developed RISPP approach, simulator, and hardware prototype (to perform power measurements for the partial run-time reconfiguration) as basis. In addition, the provided adaptivity can also be extended toward multitasking environments. As already indicated in the result section of this monograph, the RISPP approach provides significant benefits for run-time changing availability of reconfigurable fabric for a particular task. However, assigning reconfigurable fabric to a particular task is a challenging research-relevant task that may exploit the benefits that the provided RISPP architecture offers. In addition, if multiple tasks demand a reconfiguration, then the question arises, which task may perform its reconfiguration first, i.e., an extended scheduling problem for resource-management aspects of the reconfiguration port. The new "invasive computing" research initiative [Inv] investigates this matter in the scope of subproject B1 "adaptive application-specific invasive microarchitecture."

Improved design efficiency. When analyzing the utilization of the reconfigurable fabric (i.e., execution of atoms), then it becomes noticeable that it is not fully utilized, as only SI executions actually use it. Even though the SI execution typically corresponds to a large share of the application execution time (as they are designed to cover the major kernels), not all SI executions use the reconfigurable fabric, e.g., when the reconfigurations of the demanded atoms are not yet completed. Therefore, additional performance potential is available in the reconfigurable fabric that is not utilized yet. Here, a multicore RISPP system may increase the efficient utilization of the reconfigurable fabric by connecting multiple independent cores to it, i.e., the fabric can be shared among multiple cores. Ideally, when one core executes an SI using the reconfigurable fabric, the other cores execute code on their core pipeline, i.e., all cores have the impression that they may use their assigned share of the reconfigurable fabric exclusively (similar to a virtualization). Actually, when two or more cores aim to access the reconfigurable fabric at the same time, then at least one of them will face performance degradation. However, in comparison to a single-core RISPP system with multitasking, this concept may increase the performance significantly by utilizing the reconfigurable fabric efficiently. In addition, the reconfigurable fabric could be optimized toward the requirements of actual atoms and SIs, using an embedded FPGA architecture (e.g., [NvSBN08, SNBN06]). This would allow to trade-off the features of the reconfigurable fabric with its area footprint and logic delay. Certainly, the reconfigurable fabric that is used for prototyping (Xilinx Virtex-4) is overdesigned for implementing atoms, as, e.g., a noticeable amount of routing resources crosses the border of the partially reconfigurable regions, and thus they cannot be used. In addition to a fine-grained reconfigurable fabric, also coarse-grained reconfigurable elements may be used. Especially for computation on word level, they may provide better performance and area efficiency than fine-grained reconfigurable fabrics. A combination of both types would not only improve the area efficiency, but also the performance, as then both byte/sub-byte computations and word-level computations are supported with specialized fabrics. The KAHRISMA project started investigating this matter and published a first promising case study [ITI, KBS+10].

Appendix A: RISPP Simulation

To be able to investigate concepts, different algorithms, and different parameters for RISPP before implementing the prototype, a simulation environment was developed that allows a fast design-space exploration [BSH09a]. It is not the intend to explore the design space automatically, but rather an accurate and configurable simulator with a corresponding tool chain was implemented in this work that allows investigating different run-time algorithms, comparing reconfigurable processors on a fair basis, and investigating the impact of diverse architectural parameters. Figure A.1 visualizes the major components and interactions of the simulation environment as an UML class diagram. It is partitioned into three major parts: the core pipeline and run-time system (managing the application execution and reconfigurations), the SIs (representing their implementations, execution times, etc.), and the FPGA (managing the AC content, reconfiguration times, etc.).

The SIs comprise at least two molecules: one uses the atoms and one uses the core instruction set architecture (cISA). The SIs and their molecules are determined by an external XML file that defines all required information like the name of the SI, the instruction format and opcode (to be able to decode it out of the application binary), and the available molecules with the latency and atom requirements. Furthermore, the atoms contain the information on the bitstream size to determine the reconfiguration time. For different data memory ports and bit widths, different XML files exist. They are semiautomatically created as part of the tool chain, taking the actual data-flow graph of the SI (the atoms correspond to the nodes in the graph) and then scheduling it, i.e., determining a starting time for each atom. This schedule is automatically performed for all possible resource constraints (i.e., the amount of instances of a certain atom that may be used at the same time) to obtain all molecules.

The simulation of the pipeline receives the application binary and information about the cISA (for which the binary was created) as input. When the simulation starts, the cISA is automatically extended by the SI information from the XML file to be able to decode all instructions from the application binary. The cISA instructions are actually not executed by the pipeline. Instead, a branch trace (containing all taken branches in their actually executed sequence) is provided to mimic the exact application control flow. An entry in the branch trace correspond to the information at which address the corresponding branch instructions is placed (in the application binary) and to which address the control-flow branches. This information is

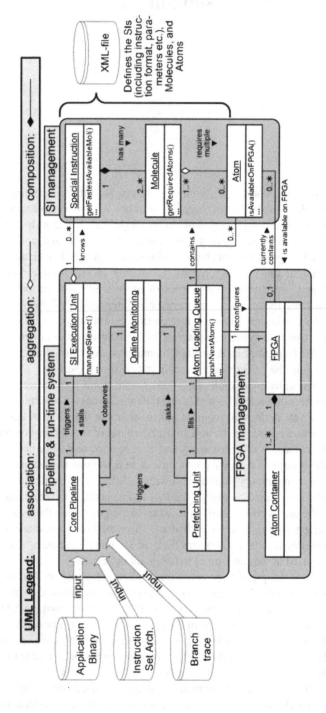

Fig. A.1 Internal composition of the design space exploration tool, showing module interactions

represented as binary values, and a run-length compression is used to reduce the size of the branch trace. To simulate the execution of the application, the instruction at address zero is provided as first instruction to the instruction fetch stage. If the stage is not stalled (described below), the subsequent instruction are provided in the next cycle, etc. This continues linearly, until the address of the first entry in the branch trace is reached (i.e., the first taken branch). Afterwards (considering delay slots, etc.), the address of the branch target is provided to the instruction fetch stage and from there on, sequentially providing the next addresses (that follow the branch target) continues until the next taken branch is reached, etc. By doing so, the exact application control flow can be simulated without simulating the semantic of the instructions to determine whether a branch instruction is taken. This allows abstracting the simulation environment from the actual cISA[1] and reduces the simulation time, as investigating architectural parameters and run-time algorithms for a particular branch trace is the goal. The branch trace is derived from an instruction set simulator (ISS); for MIPS and SPARC-V8, DLXSim [HM94] and ArchC [ARB+05] are used, respectively. It was assured that the execution in the simulator matches the actual execution. Therefore, the pipeline is modeled including its stages in the simulator and the information which instruction requires how many cycles in which pipeline stage (e.g., the *mult* instruction stalls the execution stage) is available. However, the register file and data memory accesses are not simulated, thus, although at any time it is known which instruction is currently executed in which pipeline stage, the current content of the register file or the actual data memory accesses cannot be determined. To simulate the exact data memory accesses (e.g., to attach a cache simulator), an additional data memory access trace would be needed. Currently, each load/store instruction is configured to require two cycles. When the pipeline issues these memory accesses, then the cycle time corresponds to the CPU frequency. When a molecule that executes on the atom infrastructure issues them, they correspond to the FPGA frequency. For the access to the on-chip scratchpad (see Sect. 5.3), this corresponds to the results of the hardware prototype. For accesses to the off-chip memory, this mimics the scenario that all accesses to the instruction memory and stack are cache hits. This allows investigating the differences of parameters and algorithms without affecting the results by cache effects.

When the instruction decode stage recognizes an SI, the SI execution unit is triggered, examines which molecule shall be used to execute the SI, and stalls the execute stage accordingly. When a forecast instruction is recognized, then the prefetching unit is triggered and the pipeline continues executing the next instructions. The prefetching unit implements the algorithms of the run-time system, i.e., it corresponds to the MicroBlaze in the hardware prototype. It receives the expected SI execution frequencies as input from the online monitoring, determines which atoms shall be reconfigured into which ACs, and writes this information into the

---

[1] It is currently prepared for MIPS and SPARC-V8, i.e., the SI instruction formats and opcodes for both ISAs are provided in the XML file.

atom loading queue. The online monitoring in Fig. A.1 is initialized with offline profiling data that is derived from the ISS and that is fine-tuned at run time.

The simulator creates a detailed log file containing information about the current system state, the decisions made, and intermediate results of the run-time algorithms. The planned reconfigurations are printed along with the current state of the FPGA and the information which atoms and which molecules are currently available. Furthermore, statistics on the SI executions are shown, e.g., which molecules were executed since simulation start or in the recent time. This log file can be used for analyzing the results by extracting the required information from it. In addition to the textual log file, also a visual representation was developed in the scope of the work presented in this monograph. During the simulation run, a special binary version of the log file is created, and after the simulation finished this log file is parsed by RISPPVis, the RISPP visualization. Fig. A.2 provides a first overview of the RISPPVis graphical user interface (GUI) that is based on QT and OpenGL. RISPPVis provides a timeline (see Label ①) that provides an overview of the entire application execution time and highlights forecasts (red lines) and completed atom reconfigurations (blue lines). It shows the execution flow of an H.264 video encoder. The first forecast corresponds to the motion estimation (ME), the second one to the encoding engine (EE), and the third one to the in-loop deblocking filter (LF). The fourth forecast (short after the third one) corresponds to ME of the next frame.

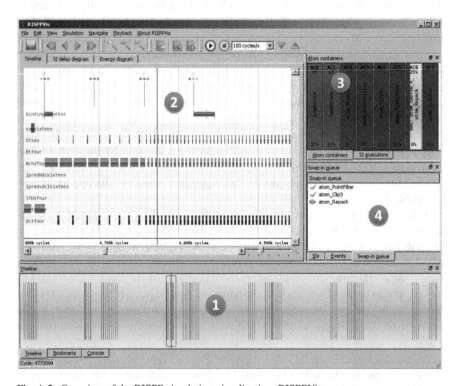

Fig. A.2 Overview of the RISPP simulation visualization: RISPPVis

Label ② in Fig. A.2 indicates the main widget that provides a detailed view of the major events and the SI executions. The events are visible in the upper half of the widget and all SIs have a dedicated row in the lower half that indicates their executions. Depending on the zoom level, multiple executions of the same SI may be summarized as one block, however, when zooming into the simulation time, then the individual SI executions are printed as individual blocks. Fig. A.3 shows such a zoom (see Label ①, the zoom slider), where the individual SI executions become apparent. In addition, the SI execution latency is written into the blocks (see Label ②). The black line below Label ② in Fig. A.2 shows the point in time for which all status information is printed. RISPPVis provides a playback feature (i.e., it can run the simulation) and all status information are refreshed accordingly. Label ③ in the figure shows the atom infrastructure that is automatically adjusted to the parameters that were used for the particular simulation, i.e., it shows the number of ACs that were used in the simulation. In the shown example, the atom infrastructure consists of two load/store units (the address generation units are not printed for clarity) and six reconfigurable atom containers (ACs). The atom that is currently loaded into an AC is written into its box. AC6 in Fig. A.2 is currently reconfigured, as indicated by the partial color filling of the AC that corresponds to the degree to that the reconfiguration is complete (25% as written below the AC number). The text in the box indicates which atom was loaded into this AC beforehand and which atom is currently

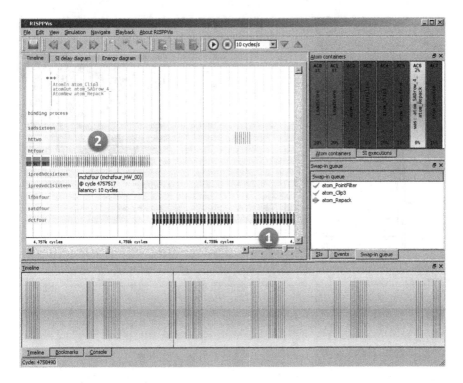

Fig. A.3 RISPPVis zooming into SI execution latency changes

reconfigured. The colors of the ACs correspond to the percentage value that is written at the bottom of each AC (0% is blue and 100% is red). These values show how often the AC was used in the recent time, i.e., it shows its utilization. Label ④ shows the swap-in queue, indicating which atoms already finished reconfiguration for the currently executed computational block and which will be reconfigured next.

In addition to this information, many different statistics about the SIs and molecules can be plotted as diagrams, e.g., showing all molecules of an SI and indicating which molecule was executed how often. Fig. A.4 provides an example for such a diagram. Here, the main widget shows the changes of the SI latencies over time, indicating the molecule selection (how fast does the implementation of a particular SI become), the atom scheduling (in which sequence are the SIs upgraded), and the atom replacement. Similar types of graphs were used when describing the atom scheduling (Fig. 4.25) and atom replacement (Fig. 4.30), and RISPPVis can create this type of graph automatically, which improves the understanding of the run-time system decisions significantly.

Furthermore, RISPPVis provides an SQL database that contains all information and events that are provided by the RISPP simulation. This allows searching for specific conditions that are then shown on the time axis in RISPPVis. The SQL database is used to implement a design rule checker to assure that certain impos-

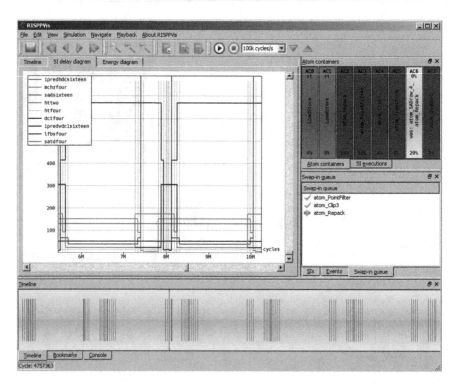

Fig. A.4 RISPPVis SI latency diagram

sible situations (e.g., two reconfigurations happening at the same time, two SIs executing at the same time, a molecule executes even though not all required atoms are available, etc.) never occur. This feature is mainly used to assure basic properties of the RISPP architecture, which is especially beneficial after changing the RISPP simulator, e.g., after adding a new feature. Fig. A.5 shows an example of the design rule check. The output is printed to the console in the RISPPVis GUI. At Label ①, an interesting event is reported: "Atom atom_Cond immediately swapped out at cycle 2,777,585." This means that an atom that just finished reconfiguration was immediately replaced, i.e., the atom was never used. This indicates potential for further improvement, for instance, it might be more efficient to avoid starting a new reconfiguration when the computational block is nearly completed. At Label ②, this situation already starts. This information reports that during a running reconfiguration, a new forecast arrives. Actually, two forecasts arrive: the first one is informing that the LF computational block completed and the second one informs that ME starts, i.e., another frame is to be encoded. Short time after this forecast, the running reconfiguration completed and leads to the message at Label ③ (please note, the messages are not sorted by cycles but they are printed one rule after the other). The report at Label ③ indicates that the number of started reconfigurations does not match the number of completed reconfigurations, because one reconfiguration was still running, when the application execution finished.

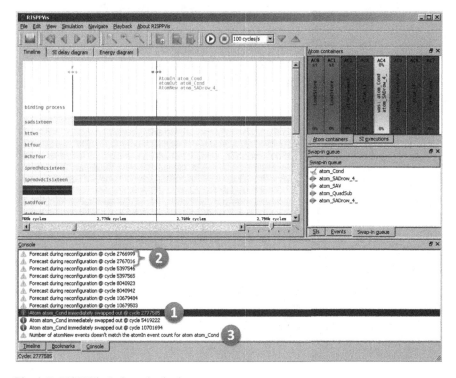

Fig. A.5 RISPPVis design rule check

Appendix B: RISPP Prototype

The hardware implementation of the RISPP prototype (see Sect. 5.5) is developed and tested for an Avnet Xilinx Virtex-4 LX160 development kit ("ADS-XLX-V4LX-DEV160-G," [Avn09]). This board comprises a Xilinx Virtex-4 XC4VLX160-FF1513 FPGA [Xil07a, Xil08b] that provides the resources shown in Table B.1.

Fig. B.1 shows an overview of the board with all connected peripherals. The FPGA is placed below the heat-sink at Label ①. The board features different types of memory, i.e., 128 MB DDR RAM (at the bottom side, basically below Label ①), 64 MB SDRAM (Label ②), two SRAM modules with 1 MB per module (Label ③), a 16 MB Flash EEPROM (Label ④) for configuring the FPGA (full bitstreams), and a 32 MB fast Flash EEPROM (Label ⑤) for configuring the ACs (partial bitstreams). The DDR RAM is connected to the MicroBlaze processor (although it currently does not require it), the SD RAM is used as a main memory for the Leon2 processor, and one SRAM module is used as a video buffer for the video IP core (to create a progressive frame out of two subsequent interleaved frames, as received from the video input). The EEPROM for the partial bitstreams is placed on a printed circuit board (PCB) that was developed in the scope of the presented work for the prototype and is described below. In addition to the EEPROM, this PCB also provides access to further peripherals, e.g., USB/UART and a 320 × 240 touch screen LCD [ELE] (Label ⑥). In addition, this PCB provides general purpose I/O that complies the Digilent PMOD[2] connector and allows connecting further peripherals, e.g., a 16 × 2 character display, SD-card slot, push buttons, and a two-axis joystick (Label ⑦). The FPGA board also comprises a breakout board that provides Mictor connectors for debugging and logging purpose (Label ⑧). The video module that is connected to the board provides video input via an S-video connector (Label ⑨) and video output via an VGA connector (Label ⑩).

To be able to store the partial bitstreams for reconfiguring the atom containers on a nonvolatile memory (i.e., they do not need to be reuploaded after power-on), a PCB was developed in the scope of the presented work to connect the fast

---

[2]http://www.digilentinc.com/Products/Catalog.cfm?NavPath=2,401&Cat=9.

Table B.1 Resources provided by the FPGA of the prototype

Device	CLB[a] array: row × column	Logic cells	Slices	Maximum distributed RAM (Kbit)	Xtreme DSP[b] slices	18 kb blocks	Maximum block RAM (Kbit)	DCMs[c]	PMCDs[d]	Total I/O banks	Maximum user I/O
XC4VLX160	192 × 88	152,064	67,584	1,056	96	288	5,184	12	8	17	960

[a] Configurable logic block
[b] Digital signal processing
[c] Digital clock managers
[d] Phase-matched clock dividers

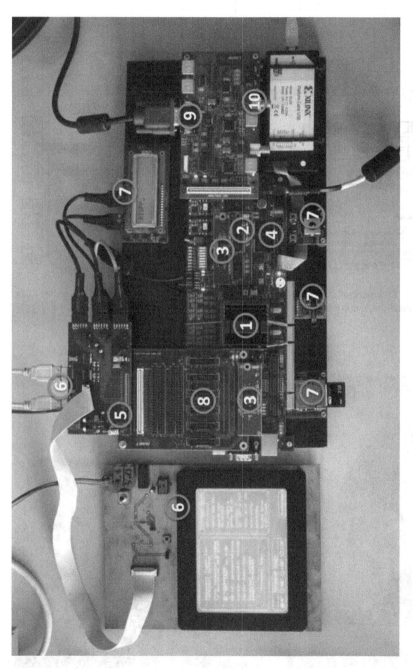

Fig. B.1 Picture of the Avnet Xilinx Virtex-4 LX160 development kit with periphery for SRAM, SDRAM, reconfiguration EEPROM, audio/video module, and (touch screen) LCDs

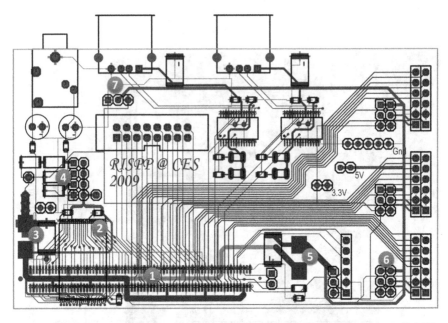

Fig. B.2 Schematic of the four layers for the PCB for EEPROM, USB, audio, touch screen LCD, and general-purpose connectors

Fig. B.3 Picture of the developed PCB

OneNAND Flash EEPROM from Samsung [Sam05] to the FPGA board. Fig. B.2
shows the four layers of the PCB and Fig. B.3 shows a picture of the PCB board
with the connected peripherals. Label ① in Fig. B.2 indicates the 140-pin AvBus
connector that establishes the connection to the FPGA board, and Label ② shows
the placement of the EEPROM right on the top of the connector. The AvBus connector provides Gnd, 3.3 V, and 5 V in addition to user-I/O pins, and four different
layers are demanded to provide the required connections to the peripherals. The
color codes for the layers in Fig. B.2 are (from top to bottom) blue (data), brown
(Vcc), green (Gnd), and red (data). Except the AvBus connector (that is placed on
the bottom side of the PCB), all components are placed on the top side. Label ③
shows the connectors for a fuse and a capacitor for 3.3 V Vcc, and at Label ④ the
Vcc connection to the EEPROM is connected to precision resistors to measure the
power consumption. Label ⑤ shows the fuse and capacitor for the 5 V Vcc which
is used for the PMOD connectors (Label ⑥) and the background light for the touch
screen LCD (Label ⑦).

Bibliography

[ABF+07] A. Ahmadinia, C. Bobda, S. P. Fekete, J. Teich, and J. C. van der Veen, "Optimal free-space management and routing-conscious dynamic placement for reconfigurable devices", *IEEE Transactions on Computers (TC)*, vol. 56, no. 5, pp. 673–680, May 2007.

[ABK+04] A. Ahmadinia, C. Bobda, D. Koch, M. Majer, and J. Teich, "Task scheduling for heterogeneous reconfigurable computers", in *Proceedings of the 17th Symposium on Integrated Circuits and System Design (SBCCI)*. ACM, September 2004, pp. 22–27.

[Aer] Aeroflex Gaisler, "Homepage of the Leon processor", http://www.gaisler.com/leonmain.html.

[Ama06] H. Amano, "A survey on dynamically reconfigurable processors", *IEICE Transaction on Communication*, vol. E89-B, no. 12, pp. 3179–3187, December 2006.

[API03] K. Atasu, L. Pozzi, and P. Ienne, "Automatic application-specific instruction-set extensions under microarchitectural constraints", in *Proceedings of the 40th Annual Conference on Design Automation (DAC)*, June 2003, pp. 256–261.

[ARB+05] R. Azevedo, S. Rigo, M. Bartholomeu, G. Araujo, C. Araujo, and E. Barros, "The ArchC architecture description language and tools", *International Journal of Parallel Programming*, vol. 33, no. 5, pp. 453–484, October 2005.

[ARC] ARC International, "ARCtangent processor", http://www.arc.com/configurables/.

[ASI] ASIP Solutions, Inc., "Homepage of ASIP Meister", http://www.asip-solutions.com/.

[Avn09] Avnet, Inc., "Avnet electronics marketing", http://www.avnetexpress.avnet.com, 2009.

[BA05] C. Bobda and A. Ahmadinia, "Dynamic interconnection of reconfigurable modules on reconfigurable devices", *IEEE Design and Test of Computers*, vol. 22, no. 5, pp. 443–451, September/October 2005.

[BAM+05] C. Bobda, A. Ahmadinia, M. Majer, J. Teich, S. Fekete, and J. van der Veen, "DyNoC: A dynamic infrastructure for communication in dynamically reconfigurable devices", in *International Conference on Field Programmable Logic and Applications (FPL)*, August 2005, pp. 153–158.

[BBKG07] F. Bouwens, M. Berekovic, A. Kanstein, and G. Gaydadjiev, "Architectural exploration of the ADRES coarse-grained reconfigurable array", in *Reconfigurable Computing: Architectures, Tools and Applications (ARC)*, March 2007, pp. 1–13.

[BCA+04] P. Biswas, V. Choudhary, K. Atasu, L. Pozzi, P. Ienne, and N. Dutt, "Introduction of local memory elements in instruction set extensions", in *Proceedings of the 41st Annual Conference on Design Automation (DAC)*, June 2004, pp. 729–734.

[Bel66]   L. A. Belady, "A study of replacement algorithms for a virtual-storage computer", *IBM Systems Journal*, vol. 5, no. 2, pp. 78–101, 1966.

[BEM+03]   V. Baumgarte, G. Ehlers, F. May, A. Nückel, M. Vorbach, and M. Weinhardt, "PACT XPP – a self-reconfigurable data processing architecture", *Journal of Supercomputing*, vol. 26, no. 2, pp. 167–184, September 2003.

[BKS04]   P. Brisk, A. Kaplan, and M. Sarrafzadeh, "Area-efficient instruction set synthesis for reconfigurable system-on-chip designs", in *Proceedings of the 41st Annual Conference on Design Automation (DAC)*, June 2004, pp. 395–400.

[BL00]   F. Barat and R. Lauwereins, "Reconfigurable instruction set processors: A survey", in *Proceedings of the 11th IEEE International Workshop on Rapid System Prototyping (RSP)*, June 2000, pp. 168–173.

[BLC07]   T. Becker, W. Luk, and P. Y. K. Cheung, "Enhancing relocatability of partial bitstreams for run-time reconfiguration", in *Proceedings of the 15th Annual IEEE Symposium on Field-Programmable Custom Computing Machines (FCCM)*, April 2007, pp. 35–44.

[BMA+05]   C. Bobda, A. Majer, A. Ahmadinia, T. Haller, A. Linarth, and J. Teich, "The Erlangen Slot Machine: A highly flexible FPGA-based reconfigurable platform", in *13th Annual IEEE Symposium on Field-Programmable Custom Computing Machines (FCCM)*, April 2005, pp. 319–320.

[BNS+04]   G. Braun, A. Nohl, W. Sheng, J. Ceng, M. Hohenauer, H. Scharwächter, R. Leupers, and H. Meyr, "A novel approach for flexible and consistent ADL-driven ASIP design", in *Proceedings of the 41st Annual Design Automation Conference (DAC)*, June 2004, pp. 717–722.

[Bob07]   C. Bobda, *Introduction to Reconfigurable Computing: Architectures, Algorithms, and Applications*. New York: Springer, June 2007.

[BSH08a]   L. Bauer, M. Shafique, and J. Henkel, "A computation- and communication-infrastructure for modular special instructions in a dynamically reconfigurable processor", in *18th International Conference on Field Programmable Logic and Applications (FPL)*, September 2008, pp. 203–208.

[BSH08b]   L. Bauer, M. Shafique, and J. Henkel, "Efficient resource utilization for an extensible processor through dynamic instruction set adaptation", *IEEE Transactions on Very Large Scale Integration Systems (TVLSI), Special Section on Application-Specific Processors*, vol. 16, no. 10, pp. 1295–1308, October 2008.

[BSH08c]   L. Bauer, M. Shafique, and J. Henkel, "Run-time instruction set selection in a transmutable embedded processor", in *Proceedings of the 45th Annual Conference on Design Automation (DAC)*, June 2008, pp. 56–61, Received a "European Network of Excellence on High Performance and Embedded Architecture and Compilation" HiPEAC PAPER AWARD.

[BSH09a]   L. Bauer, M. Shafique, and J. Henkel, "Cross-architectural design space exploration tool for reconfigurable processors", in *Proceedings of the 12th Conference on Design, Automation and Test in Europe (DATE)*, April 2009, pp. 958–963.

[BSH09b]   L. Bauer, M. Shafique, and J. Henkel, "Mindeg: A performance-guided replacement policy for run-time reconfigurable accelerators", in *IEEE International Conference on Hardware-Software Codesign and System Synthesis (CODES+ISSS)*, October 2009, pp. 335–342.

[BSKH07]   L. Bauer, M. Shafique, S. Kramer, and J. Henkel, "RISPP: Rotating instruction set processing platform", in *Proceedings of the 44th Annual Conference on Design Automation (DAC)*, June 2007, pp. 791–796.

[BSKH08]   L. Bauer, M. Shafique, S. Kreutz, and J. Henkel, "Run-time system for an extensible embedded processor with dynamic instruction set", in *Proceedings of the Conference on Design, Automation and Test in Europe (DATE)*, March 2008, pp. 752–757, Received the IEEE/ACM Design Automation and Test, DATE 2008 BEST PAPER AWARD.

[BSTH07]   L. Bauer, M. Shafique, D. Teufel, and J. Henkel, "A self-adaptive extensible embedded processor", in *First International Conference on Self-Adaptive and Self-Organizing Systems (SASO)*, July 2007, pp. 344–347.

[CAK+07] A. Chattopadhyay, W. Ahmed, K. Karuri, D. Kammler, R. Leupers, G. Ascheid, and H. Meyr, "Design space exploration of partially re-configurable embedded processors", in *Proceedings of the Conference on Design, Automation and Test in Europe (DATE)*, April 2007, pp. 319–324.

[CBC+05] N. Clark, J. Blome, M. Chu, S. Mahlke, S. Biles, and K. Flautner, "An architecture framework for transparent instruction set customization in embedded processors", in *Proceedings of the 32nd International Symposium on Computer Architecture (ISCA)*, June 2005, pp. 272–283.

[CC01] J. E. Carrillo and P. Chow, "The effect of reconfigurable units in superscalar processors", in *Proceedings of the ACM/SIGDA Eighth International Symposium on Field Programmable Gate Arrays (FPGA)*, February 2001, pp. 141–150.

[CEL+03] S. Ciricescu, R. Essick, B. Lucas, P. May, K. Moat, J. Norris, M. Schuette, and A. Saidi, "The reconfigurable streaming vector processor (RSVPTM)", in *Proceedings of the 36th Annual IEEE/ACM International Symposium on Microarchitecture (MICRO)*, December 2003, pp. 141–150.

[CH02] K. Compton and S. Hauck, "Reconfigurable computing: A survey of systems and software", *ACM Computing Surveys (CSUR)*, vol. 34, no. 2, pp. 171–210, June 2002.

[CHP03] N. Cheung, J. Henkel, and S. Parameswaran, "Rapid configuration and instruction selection for an ASIP: A case study", in *IEEE/ACM Proceedings of Design Automation and Test in Europe (DATE)*, March 2003, pp. 802–807.

[CKP+04] N. Clark, M. Kudlur, H. Park, S. Mahlke, and K. Flautner, "Application-specific processing on a general-purpose core via transparent instruction set customization", in *Proceedings of the 37th Annual IEEE/ACM International Symposium on Microarchitecture (MICRO)*, December 2004, pp. 30–40.

[CLC+02] K. Compton, Z. Li, J. Cooley, S. Knol, and S. Hauck, "Configuration relocation and defragmentation for run-time reconfigurable computing", *IEEE Transactions on Very Large Scale Integration Systems (TVLSI)*, vol. 10, no. 3, pp. 209–220, June 2002.

[CMZS07] C. Claus, F. Müller, J. Zeppenfeld, and W. Stechele, "A new framework to accelerate Virtex-II Pro dynamic partial self-reconfiguration", in *IEEE International Parallel and Distributed Processing Symposium (IPDPS)*, March 2007, pp. 1–7.

[CoW] CoWare Inc., "LISATek", http://www.coware.com/.

[CPH04] N. Cheung, S. Parameswarani, and J. Henkel, "A quantitative study and estimation models for extensible instructions in embedded processors", in *Proceedings of the IEEE/ACM International Conference on Computer-Aided Design (ICCAD)*, November 2004, pp. 183–189.

[CZM03] N. Clark, H. Zhong, and S. Mahlke, "Processor acceleration through automated instruction set customization", in *Proceedings of the 36th Annual IEEE/ACM International Symposium on Microarchitecture (MICRO)*, December 2003, pp. 129–140.

[Dal99] M. Dales, "The Proteus processor – a conventional CPU with reconfigurable functionality", in *Proceedings of the Ninth International Workshop on Field-Programmable Logic and Applications (FPL)*, August 1999, pp. 431–437.

[Dal03] M. Dales, "Managing a reconfigurable processor in a general purpose workstation environment", in *Design, Automation and Test in Europe Conference and Exhibition (DATE)*, March 2003, pp. 980–985.

[EL09] A. Ehliar and D. Liu, "An ASIC perspective on FPGA optimizations", in *Proceedings of the 19th International Conference on Field-Programmable Logic and Applications (FPL)*, August/September 2009, pp. 218–223.

[ELE] ELECTRONIC ASSEMBLY GmbH, "eDIP embedded LCD-display", http://www.lcd-module.com/eng/pdf/grafik/edip320-8e.pdf.

[ESS+96] H. ElGindy, A. K. Somani, H. Schroder, H. Schmeck, and A. Spray, "RMB – a reconfigurable multiple bus network", in *Proceedings of the Second IEEE Symposium on High-Performance Computer Architecture (HPCA)*, February 1996, pp. 108–117.

[FKPM06]   K. Fan, M. Kudlur, H. Park, and S. Mahlke, "Increasing hardware efficiency with multifunction loop accelerators", in *Proceedings of the Fourth International Conference on Hardware/Software Codesign and System Synthesis (CODES+ISSS)*, October 2006, pp. 276–281.

[GJ90]   M. R. Garey and D. S. Johnson, *Computers and Intractability; A Guide to the Theory of NP-Completeness*. New York: W. H. Freeman & Co., 1990.

[GRE+01]   M. Guthaus, J. Ringenberg, D. Ernst, T. Austin, T. Mudge, and R. Brown, "MiBench: A free, commercially representative embedded benchmark suite", in *Annual IEEE International Workshop Workload Characterization (WWC)*, December 2001, pp. 3–14.

[GSB+00]   S. C. Goldstein, H. Schmit, M. Budiu, S. Cadambi, M. Moe, and R. R. Taylor, "Piperench: A reconfigurable architecture and compiler", *Computer*, vol. 33, no. 4, pp. 70–77, April 2000.

[GVPR04]   B. Griese, E. Vonnahme, M. Porrmann, and U. Rückert, "Hardware support for dynamic reconfiguration in reconfigurable SoC architectures", in *Proceedings of the 14th International Conference on Field-Programmable Logic and Applications (FPL)*, August/September 2004, pp. 842–846.

[Har01]   R. Hartenstein, "A decade of reconfigurable computing: A visionary retrospective", in *Proceedings of the Conference on Design, Automation and Test in Europe (DATE)*, March 2001, pp. 642–649.

[Hen03]   J. Henkel, "Closing the SoC design gap", *Computer*, vol. 36, no. 9, pp. 119–121, September 2003.

[HFHK97]   S. Hauck, T. W. Fry, M. M. Hosler, and J. P. Kao, "The Chimaera reconfigurable functional unit", in *Proceedings of the Fifth IEEE Symposium on FPGA-Based Custom Computing Machines (FCCM)*, April 1997, pp. 87–96.

[HGG+99]   A. Halambi, P. Grun, V. Ganesh, A. Khare, N. Dutt, and A. Nicolau, "EXPRESSION: A language for architecture exploration through compiler/simulator retargetability", in *Proceedings of the Conference on Design, Automation and Test in Europe (DATE)*, March 1999, pp. 485–490.

[HKN+01]   A. Hoffmann, T. Kogel, A. Nohl, G. Braun, O. Schliebusch, O. Wahlen, A. Wieferink, and H. Meyr, "A novel methodology for the design of application-specific instruction-set processors (ASIPs) using a machine description language", *IEEE Transactions on Computer-Aided Design of Integrated Circuits and Systems (TCAD)*, vol. 20, no. 11, pp. 1338–1354, November 2001.

[HM94]   L. B. Hostetler and B. Mirtich, "DLXsim – a simulator for DLX", http://www.heather.cs.ucdavis.edu/~matloff/DLX/Report.html, 1994.

[HM09]   H. P. Huynh and T. Mitra, "Runtime adaptive extensible embedded processors – a survey", in *Proceedings of the Ninth International Workshop on Embedded Computer Systems: Architectures, Modeling, and Simulation (SAMOS)*, July 2009, pp. 215–225.

[HMW04]   L. He, T. Mitra, and W.-F. Wong, "Configuration bitstream compression for dynamically reconfigurable FPGAs", in *Proceedings of the IEEE/ACM International Conference on Computer-Aided Design (ICCAD)*, November 2004, pp. 766–773.

[HP96]   J. L. Hennessy and D. A. Patterson, *Computer Architecture – A Quantitative Approach*, 2nd ed. San Francisco: Morgan Kaufmann, 1996.

[Hro01]   J. Hromkovic, *Algorithmics for Hard Problems: Introduction to Combinatorial Optimization, Randomization, Approximation, and Heuristics*. New York: Springer, 2001.

[HSKB06]   M. Hübner, C. Schuck, M. Kühnle, and J. Becker, "New 2-dimensional partial dynamic reconfiguration techniques for real-time adaptive microelectronic circuits", in *Proceedings of the IEEE Computer Society Annual Symposium on Emerging VLSI Technologies and Architectures (ISVLSI)*, August/September 2006, pp. 97–102.

[HSM03]   P. Heysters, G. Smit, and E. Molenkamp, "A flexible and energy-efficient coarse-grained reconfigurable architecture for mobile systems", *Journal of Supercomputing*, vol. 26, no. 3, pp. 283–308, November 2003.

[HSM07]   H. P. Huynh, J. E. Sim, and T. Mitra, "An efficient framework for dynamic recon-figuration of instruction-set customization", in *Proceedings of the International Conference on Compilers, Architecture, and Synthesis for Embedded Systems (CASES)*, September/October 2007, pp. 135–144.

[HUWB04]   M. Hübner, M. Ullmann, F. Weissel, and J. Becker, "Real-time configuration code decompression for dynamic FPGA self-reconfiguration", in *Proceedings of the 18th International Parallel and Distributed Processing Symposium (IPDPS)*, April 2004, pp. 138–143.

[IHT+00]   M. Itoh, S. Higaki, Y. Takeuchi, A. Kitajima, M. Imai, J. Sato, and A. Shiomi, "PEAS-III: An ASIP design environment", in *International Conference on Computer Design (ICCD)*, September 2000, pp. 430–436.

[Inv]   InvasIC, "Homepage of the Invasive Computing initiative (InvasIC)", http://www.invasic.de/.

[ISS]   ISS LISA Team, "LISA – language for instruction set architectures", http://www.iss.rwth-aachen.de/Projekte/Tools/lisa/index.html.

[ITI]   ITIV & CES, "KAHRISMA: KArlsruhe's Hypermorphic Reconfigurable-Instruction-Set Multi-grained-Array processor", http://www.kahrisma.org/.

[ITU05]   ITU-T Rec. H. 264 and ISO/IEC 14496-10:2005 (E) (MPEG-4 AVC), "Advanced video coding for generic audiovisual services", 2005.

[JC99]   J. A. Jacob and P. Chow, "Memory interfacing and instruction specification for reconfigurable processors", in *Proceedings of the ACM/SIGDA Seventh International Symposium on Field Programmable Gate Arrays (FPGA)*, February 1999, pp. 145–154.

[KBS+10]   R. König, L. Bauer, T. Stripf, M. Shafique, W. Ahmed, J. Becker, and J. Henkel, "KAHRISMA: A novel hypermorphic reconfigurable-instruction-set multi-grained-array architecture", in *Proceedings of the Conference on Design, Automation and Test in Europe (DATE)*, March 2010, pp. 819–824.

[KLPR05]   H. Kalte, G. Lee, M. Porrmann, and U. Rückert, "REPLICA: A bitstream manipu-lation filter for module relocation in partial reconfigurable systems", in *Reconfigurable Architectures Workshop (RAW), Proceedings of the 19th IEEE International Parallel and Distributed Processing Symposium (IPDPS)*, April 2005, p. 151.2.

[KMN02]   K. Keutzer, S. Malik, and A. R. Newton, "From ASIC to ASIP: The next design discontinuity", in *International Conference on Computer Design (ICCD), IEEE Computer Society*, September 2002, pp. 84–90.

[KMTI03]   S. Kobayashi, K. Mita, Y. Takeuchi, and M. Imai, "Rapid prototyping of JPEG encoder using the ASIP development system: PEAS-III", in *Proceedings of the International Conference on Multimedia and Expo (ICME)*, April 2003, pp. 149–152.

[KT04]   D. Koch and J. Teich, "Platform-independent methodology for partial reconfigura-tion", in *Proceedings of the First Conference on Computing Frontiers (CF)*, April 2004, pp. 398–403.

[LBM+06]   P. Lysaght, B. Blodget, J. Mason, J. Young, and B. Bridgford, "Enhanced archi-tectures, design methodologies and CAD tools for dynamic reconfiguration of Xilinx FPGAs", in *Proceedings of the 16th International Conference on Field-Programmable Logic and Applications (FPL)*, August 2006, pp. 1–6.

[LH01]   Z. Li and S. Hauck, "Configuration compression for Virtex FPGAs", in *Proceedings of the Ninth Annual IEEE Symposium on Field-Programmable Custom Computing Machines (FCCM)*, April/May 2001, pp. 147–159.

[LH02]   Z. Li and S. Hauck, "Configuration prefetching techniques for partial reconfigu-rable coprocessor with relocation and defragmentation", in *Proceedings of Eighth*

International Symposium on Field Programmable Gate Arrays (FPGA), February 2002, pp. 187–195.

[LLC06]   A. Lopez-Lagunas and S. M. Chai, "Compiler manipulation of stream descriptors for data access optimization", in *Proceedings of the International Conference Workshops on Parallel Processing (ICPPW)*, August 2006, pp. 337–344.

[LP07]   E. Lübbers and M. Platzner, "ReconOS: An RTOS supporting hard- and software threads", in *International Conference on Field Programmable Logic and Applications (FPL)*, August 2007, pp. 441–446.

[LPMS97]   C. Lee, M. Potkonjak, and W. H. Mangione-Smith, "MediaBench: A tool for evaluating and synthesizing multimedia and communications systems", in *Proceedings of the 36th Annual IEEE/ACM International Symposium on Microarchitecture (MICRO)*, December 1997, pp. 330–335.

[LSV06]   R. Lysecky, G. Stitt, and F. Vahid, "Warp processors", *ACM Transactions on Design Automation of Electronic Systems (TODAES)*, vol. 11, no. 3, pp. 659–681, June 2006.

[LTC+03]   A. Lodi, M. Toma, F. Campi, A. Cappelli, R. Canegallo, and R. Guerrieri, "A VLIW processor with reconfigurable instruction set for embedded applications", *IEEE Journal of Solid-State Circuits (JSSC)*, vol. 38, no. 11, pp. 1876–1886, November 2003.

[LV04]   R. Lysecky and F. Vahid, "A configurable logic architecture for dynamic hardware/software partitioning", in *Proceedings of the Conference on Design, Automation and Test in Europe (DATE)*, February 2004, p. 10480.

[MBTC06]   C. Mucci, M. Bocchi, M. Toma, and F. Campi, "A case-study on multimedia applications for the XiRisc reconfigurable processor", in *Proceedings of the International Symposium on Circuits and Systems (ISCAS)*, May 2006, pp. 4859–4862.

[MLV+05]   B. Mei, A. Lambrechts, D. Verkest, J.-Y. Mignolet, and R. Lauwereins, "Architecture exploration for a reconfigurable architecture template", *IEEE Design and Test on Computers*, vol. 22, no. 2, pp. 90–101, March/April 2005.

[MO99]   T. Miyamori and K. Olukotun, "REMARC: Reconfigurable multimedia array coprocessor", *IEICE Transactions on Information and Systems*, vol. E82-D, no. 2, pp. 389–397, February 1999.

[MSCL06]   T. Mak, P. Sedcole, P. Cheung, and W. Luk, "On-FPGA communication architectures and design factors", in *International Conference on Field Programmable Logic and Applications (FPL)*, August 2006, pp. 1–8.

[MT90]   S. Martello and P. Toth, *Knapsack Problems: Algorithms and Computer Implementations*. New York: Wiley, 1990.

[MTAB07]   M. Majer, J. Teich, A. Ahmadinia, and C. Bobda, "The Erlangen Slot Machine: A dynamically reconfigurable FPGA-based computer", *Journal of VLSI Signal Processing Systems*, vol. 47, no. 1, pp. 15–31, April 2007.

[MVV+02]   B. Mei, S. Vernalde, D. Verkest, H. D. Man, and R. Lauwereins, "DRESC: A retargetable compiler for coarse-grained reconfigurable architectures", in *Proceedings of the IEEE International Conference on Field-Programmable Technology (FPT)*, December 2002, pp. 166–173.

[MVV+03]   B. Mei, S. Vernalde, D. Verkest, H. D. Man, and R. Lauwereins, "ADRES: An architecture with tightly coupled VLIW processor and coarse-grained reconfigurable matrix", in *Proceedings of the 13th International Conference on Field-Programmable Logic and Applications (FPL)*, September 2003, pp. 61–70.

[NvSBN08] B. Neumann, T. von Sydow, H. Blume, and T. G. Noll, "Design flow for embedded FPGAs based on a flexible architecture template", in *Proceedings of the Conference on Design, Automation and Test in Europe (DATE)*, March 2008, pp. 56–61.

[OBL+04]   J. Ostermann, J. Bormans, P. List, D. Marpe, M. Narroschke, F. Pereira, T. Stockhammer, and T. Wedi, "Video coding with H.264/AVC: Tools, performance, and complexity", *IEEE Circuits and Systems Magazine*, vol. 4, no. 1, pp. 7–28, 2004.

[PBV06]   E. M. Panainte, K. Bertels, and S. Vassiliadis, "Compiler-driven FPGA-area allocation for reconfigurable computing", in *Proceedings of the Conference on Design, Automation and Test in Europe (DATE)*, March 2006, pp. 369–374.

[PBV07]   E. M. Panainte, K. Bertels, and S. Vassiliadis, "The Molen compiler for reconfigurable processors", *ACM Transactions on Embedded Computing Systems (TECS)*, vol. 6, no. 1, February 2007.

[PKAM09] T. Pionteck, R. Koch, C. Albrecht, and E. Mähle, "A design technique for adapting number and boundaries of reconfigurable modules at runtime", *International Journal of Reconfigurable Computing (IJRC)*, vol. 2009, no. 942930, 2009.

[Rec]   Recore Systems, "Montium tile processor", http://www.recoresystems.com/.

[Ric]   Richard M. Stallman and the GCC Developer Community, "Using the GNU compiler collection", http://www.gcc.gnu.org/onlinedocs/gcc-4.4.2/gcc.pdf.

[RM04]   J. Resano and D. Mozos, "Specific scheduling support to minimize the reconfiguration overhead of dynamically reconfigurable hardware", in *Proceedings of the 41st Annual Design Automation Conference (DAC)*, June 2004, pp. 119–124.

[RS94]   R. Razdan and M. D. Smith, "A high-performance microarchitecture with hardware-programmable functional units", in *Proceedings of the 27th Annual International Symposium on Microarchitecture (MICRO)*, November/December 1994, pp. 172–180.

[Sam05]   Samsung Electronics Company, Ltd, "OneNAND specification", http://www.origin2.samsung.com/global/system/business/semiconductor/product/2007/6/11/OneNAND/256Mbit/KFG5616Q1A/ds_kfg5616x1a_66mhz_rev12.pdf, 2005.

[SB98]   R. S. Sutton and A. G. Barto, *Reinforcement Learning: An Introduction*. Book Ed. Cambridge: MIT Press, 1998.

[SBH09a] M. Shafique, L. Bauer, and J. Henkel, "Optimizing the H.264/AVC video encoder application structure for reconfigurable and application-specific platforms", *Journal of Signal Processing Systems (JSPS)*, vol. 60, no. 2, pp. 183–210, 2009.

[SBH09b] M. Shafique, L. Bauer, and J. Henkel, "REMiS: Run-time energy minimization scheme in a reconfigurable processor with dynamic power-gated instruction set", in *27th International Conference on Computer-Aided Design (ICCAD)*, November 2009, pp. 55–62.

[SGS98]   S. Sawitzki, A. Gratz, and R. G. Spallek, "CoMPARE: A simple reconfigurable processor architecture exploiting instruction level parallelism", in *Fifth Australasian Conference on Parallel and Real-Time Systems (PART)*, September 1998, pp. 213–224.

[SKHB08] C. Schuck, M. Kühnle, M. Hübner, and J. Becker, "A framework for dynamic 2D placement on FPGAs", in *Proceedings of 18th International Parallel and Distributed Processing Symposium (IPDPS)*, April 2008, pp. 1–7.

[SNBN06] T. V. Sydow, B. Neumann, H. Blume, and T. G. Noll, "Quantitative analysis of embedded FPGA-architectures for arithmetic", in *Proceedings of the IEEE 17th International Conference on Application-Specific Systems, Architectures and Processors (ASAP)*, September 2006, pp. 125–131.

[SPA]   SPARC International, Inc., "The SPARC architecture manual, version 8", http://www.sparc.org/specificationsDocuments.html#V8; http://www.gaisler.com/doc/sparcv8.pdf.

[SRRJ03] F. Sun, S. Ravi, A. Raghunathan, and N. K. Jha, "A scalable application-specific processor synthesis methodology", in *Proceedings of the IEEE/ACM International Conference on Computer-Aided Design (ICCAD)*, November 2003, pp. 283–290.

[SRRJ04] F. Sun, S. Ravi, A. Raghunathan, and N. Jha, "Custom-instruction synthesis for extensible-processor platforms", *IEEE Transactions on Computer-Aided Design of Integrated Circuits and Systems (TCAD)*, vol. 23, no. 2, pp. 216–228, February 2004.

[SSC03]   T. Sherwood, S. Sair, and B. Calder, "Phase tracking and prediction", in *30th Annual International Symposium on Computer Architecture (ISCA)*, June 2003, pp. 336–347.

[Sut88]   R. S. Sutton, "Learning to predict by the methods of temporal differences", *Machine Learning*, vol. 3, no. 1, pp. 9–44, August 1988.

[SWP04]   C. Steiger, H. Walder, and M. Platzner, "Operating systems for reconfigurable embedded platforms: Online scheduling of real-time tasks", *IEEE Transactions on Computers (TC)*, vol. 53, no. 11, pp. 1393–1407, November 2004.

[Tan07]   A. S. Tanenbaum, *Modern Operating Systems*. New Jersey: Prentice Hall, 2007.

[Tar]   Target Compiler Technologies NV, "Target compiler", http://www.retarget.com/.

[TB05]   A. Thomas and J. Becker, "Multi-grained reconfigurable datapath structures for online-adaptive reconfigurable hardware architectures", in *Proceedings of the IEEE Computer Society Annual Symposium on VLSI (ISVLSI)*, May 2005, pp. 118–123.

[TCW+05]   T. Todman, G. Constantinides, S. Wilton, O. Mencer, W. Luk, and P. Cheung, "Reconfigurable computing: Architectures and design methods", *IEEE Proceedings Computers and Digital Techniques*, vol. 152, no. 2, pp. 193–207, March 2005.

[Tec06]   P. X. Technologies, "XPP-III processor overview (white paper), v2.0.1", http://www.pactxpp.com/main/download/XPP-III_overview_WP.pdf, July 2006.

[Tei97]   J. Teich, *Digitale Hardware/Software-Systeme: Synthese und Optimierung*. Heidelberg: Springer, 1997.

[Tena]   Tensilica Inc., "Tensilica: Customizable processor cores for the dataplane", http://www.tensilica.com/.

[Tenb]   Tensilica Inc., "Xtensa LX2 I/O bandwidth", http://www.tensilica.com/products/io_bandwidth.htm.

[TKB+07]   F. Thoma, M. Kuhnle, P. Bonnot, E. Panainte, K. Bertels, S. Goller, A. Schneider, S. Guyetant, E. Schuler, K. Muller-Glaser, and J. Becker, "MORPHEUS: Heterogeneous reconfigurable computing", in *International Conference on Field Programmable Logic and Applications (FPL)*, August 2007, pp. 409–414.

[UHGB04a]   M. Ullmann, M. Hübner, B. Grimm, and J. Becker, "An FPGA run-time system for dynamical on-demand reconfiguration", in *Proceedings of 18th International Parallel and Distributed Processing Symposium (IPDPS)*, April 2004, pp. 135–142.

[UHGB04b]   M. Ullmann, M. Hübner, B. Grimm, and J. Becker, "On-demand FPGA run-time system for dynamical reconfiguration with adaptive priorities", in *International Conference on Field Programmable Logic and Applications (FPL)*, August 2004, pp. 454–463.

[VBI07]   A. K. Verma, P. Brisk, and P. Ienne, "Rethinking custom ISE identification: A new processor-agnostic method", in *Proceedings of the International Conference on Compilers, Architecture, and Synthesis for Embedded Systems (CASES)*, September/October 2007, pp. 125–134.

[VS07]   S. Vassiliadis and D. Soudris, *Fine- and Coarse-Grain Reconfigurable Computing*. New York: Springer, 2007.

[VWG+04]   S. Vassiliadis, S. Wong, G. Gaydadjiev, K. Bertels, G. Kuzmanov, and E. Panainte, "The MOLEN polymorphic processor", *IEEE Transactions on Computers (TC)*, vol. 53, no. 11, pp. 1363–1375, November 2004.

[WC96]   R. Wittig and P. Chow, "OneChip: An FPGA processor with reconfigurable logic", in *IEEE Symposium on FPGAs for Custom Computing Machines*, April 1996, pp. 126–135.

[WH95]   M. Wirthlin and B. Hutchings, "A dynamic instruction set computer", in *Proceedings of the IEEE Symposium on FPGAs for Custom Computing Machines (FCCM)*, April 1995, pp. 99–107.

[WSP03]   H. Walder, C. Steiger, and M. Platzner, "Fast online task placement on FPGAs: Free space partitioning and 2D-hashing", in *Reconfigurable Architectures*

Workshop (RAW), Proceedings of the International Parallel and Distributed Processing Symposium (IPDPS), April 2003, pp. 178–185.

[WTS+97]   E. Waingold, M. Taylor, D. Srikrishna, V. Sarkar, W. Lee, V. Lee, J. Kim, M. Frank, P. Finch, R. Barua, S. Babb, J. Amarasinghe, and A. Agarwal, "Baring it all to software: Raw machines", *Computer*, vol. 30, no. 9, pp. 86–93, September 1997.

[Xil05a]   Xilinx, Inc., "Using look-up tables as distributed RAM in Spartan-3 generation FPGAs, v2.0", http://www.xilinx.com/support/documentation/application_notes/xapp464.pdf, March 2005.

[Xil05b]   Xilinx, Inc., "Xilinx development system: Partial reconfiguration", http://www.toolbox.xilinx.com/docsan/xilinx8/de/dev/partial.pdf, April 2005.

[Xil07a]   Xilinx Inc., "Virtex-4 family overview, v3.0", http://www.xilinx.com/support/documentation/data_sheets/ds112.pdf, September 2007.

[Xil07b]   Xilinx, Inc., "Virtex-II platform FPGA user guide, v2.2", http://www.xilinx.com/support/documentation/user_guides/ug002.pdf, November 2007.

[Xil08a]   Xilinx, Inc., "Partial reconfiguration – PlanAhead flow FAQ", http://www.xilinx.com/support/answers/25018.htm, 2008.

[Xil08b]   Xilinx Inc., "Virtex-4 FPGA user guide, v2.6", http://www.xilinx.com/support/documentation/user_guides/ug070.pdf, December 2008.

[Xil09a]   Xilinx, Inc., "Fast simplex link (FSL) bus, v2.11b", http://www.xilinx.com/support/documentation/ip_documentation/fsl_v20.pdf, June 2009.

[Xil09b]   Xilinx, Inc., "MicroBlaze soft processor core", http://www.xilinx.com/tools/microblaze.htm, 2009.

[Xil09c]   Xilinx, Inc., "Virtex-4 FPGA configuration user guide, v1.11", http://www.xilinx.com/support/documentation/user_guides/ug071.pdf, June 2009.

[YMHB00]   Z. A. Ye, A. Moshovos, S. Hauck, and P. Banerjee, "CHIMAERA: A high-performance architecture with a tightly-coupled reconfigurable functional unit", in *Proceedings of the 27th Annual International Symposium on Computer Architecture (ISCA)*, June 2000, pp. 225–235.

Index